Science and Technology Concepts–Secondary™

Experimenting with Mixtures, Compounds, and Elements

Student Guide

National Science Resources Center

The National Science Resources Center (NSRC) is operated by the Smithsonian Institution to improve the teaching of science in the nation's schools. The NSRC disseminates information about exemplary teaching resources, develops curriculum materials, and conducts outreach programs of leadership development and technical assistance to help school districts implement inquiry-centered science programs.

Smithsonian Institution

The Smithsonian Institution was created by an act of Congress in 1846 "for the increase and diffusion of knowledge..." This independent federal establishment is the world's largest museum complex and is responsible for public and scholarly activities, exhibitions, and research projects nationwide and overseas. Among the objectives of the Smithsonian is the application of its unique resources to enhance elementary and secondary education.

STC Program™ Project Sponsors

National Science Foundation

Bristol-Meyers Squibb Foundation

Dow Chemical Company

DuPont Company

Hewlett-Packard Company

The Robert Wood Johnson Foundation

Carolina Biological Supply Company

Science and Technology Concepts–Secondary™

Experimenting with Mixtures, Compounds, and Elements

Student Guide

The STC *Program*™

Smithsonian Institution
National Science Resources Center

www.carolinacurriculum.com

Published by Carolina Biological Supply Company
Burlington, North Carolina

NOTICE This material is based upon work supported by the National Science Foundation under Grant No. ESI-9618091. Any opinions, findings, and conclusions or recommendations expressed in this material are those of the authors and do not necessarily reflect views of the National Science Foundation or the Smithsonian Institution.

This project was supported, in part, by the **National Science Foundation**.
Opinions expressed are those of the authors and not necessarily those of the foundation.

ISBN 978-1-4350-0690-4

Published by Carolina Biological Supply Company, 2700 York Road, Burlington, NC 27215.
Call toll free 1-800-334-5551.

Science and Technology Concepts—Secondary™
Experimenting with Mixtures, Compounds, and Elements

The following revision was based on the STC/MS™ module *Properties of Matter.*

Developer
Kitty Lou Smith

Scientific Reviewer
Jerry A. Bell
Senior Scientist, Education Division
American Chemical Society

Illustrator
Susie Duckworth

Photo Research
Jane Martin
Elizabeth Klemick

National Science Resources Center Staff

Executive Director
Sally Goetz Shuler

Program Specialist/Revision Manager
Elizabeth Klemick

Contractor, Curriculum Research and Development
Devin Reese

Publications Graphics Specialist
Heidi M. Kupke

Carolina Biological Supply Company Staff

Director of Product and Development
Cindy Morgan

Product Manager, The STC Program™
Jack Ashton

Curriculum Editors
Lauren Goldsmith
Gary Metheny

Managing Editor, Curriculum Materials
Cindy Vines Bright

Publications Designers
Trey Foster
Charles Thacker
Greg Willette

Science and Technology Concepts for Middle Schools™
Properties of Matter
Original Publication

Module Development Staff

Developer/Writer
David Marsland

Science Advisor
Michael John Tinnesand
Head, K-12 Science
American Chemical Society

Contributing Writers
Linda Harteker
Robert Taylor

Illustrator
Max-Karl Winkler

STC/MS™ Project Staff

Principal Investigators
Douglas Lapp, Executive Director, NSRC
Sally Goetz Schuler, Deputy Director, NSRC

Project Director
Kitty Lou Smith

Curriculum Developers
David Marsland
Henry Milne
Carol O'Donnell
Dane J. Toler

Illustration Coordinator
Max-Karl Winkler

Photo Editor
Janice Campion

Graphic Designer
Heidi M. Kupke

STC/MS™ Project Advisors

Judy Barille, Chemistry Teacher, Fairfax County, Virginia, Public Schools

Steve Christiansen, Science Instructional Specialist, Montgomery County, Maryland, Public Schools

John Collette, Director of Scientific Affairs (retired), DuPont Company

Cristine Creange, Biology Teacher, Farifax County, Virginia, Public Schools

Robert DeHaan, Professor of Physiology, Emory University Medical School

Stan Doore, Meteorologist (retired), National Oceanic and Atmospheric Administration, National Weather Service

Ann Dorr, Earth Science Teacher (retired), Fairfax County, Virginia, Public Schools; Board Member, Minerals Information Institute

Yvonne Forsberg, Physiologist, Howard Hughes Medical Center

John Gastineau, Physics Consultant, Vernier Corporation

Patricia Hagan, Science Project Specialist, Montgomery County, Maryland, Public Schools

Alfred Hall, Staff Associate, Eisenhower Regional Consortium at Appalachian Educational Laboratory

Connie Hames, Geology Teacher, Stafford County, Virginia, Public Schools

Jayne Hart, Professor of Biology, George Mason University

Michelle Kipke, Director, Forum on Adolescence, Institute of Medicine

John Layman, Professor Emeritus of Physics, University of Maryland

Thomas Liao, Professor of Engineering, State University of New York at Stony Brook

Ian MacGregor, Senior Science Advisor, Geoscience Education, National Science Foundation

Ed Mathews, Physical Science Teacher, Fairfax County, Virginia, Public Schools

Ted Maxwell, Geomorphologist, National Air and Space Museum, Smithsonian Institution

Tom O'Haver, Professor of Chemistry/Science Education, University of Maryland

Robert Ridky, Professor of Geology, University of Maryland

Mary Alice Robinson, Science Teacher, Stafford County, Virginia, Public Schools

Bob Ryan, Chief Meteorologist, WRC Channel 4, Washington, D.C.

Michael John Tinnesand, Head, K-12 Science, American Chemical Society

Grant Woodwell, Professor of Geology, Mary Washington College

Thomas Wright, Geologist, National Museum of Natural History, Smithsonian Institution; U.S. Geological Survey (emeritus)

Acknowledgments

The National Science Resources Center gratefully acknowledges the following individuals and school systems for their assistance with the national field-testing of *Properties of Matter:*

Delaware

Site Coordinator

Suzanne Curry, Director of Curriculum
Office of Climate and Standards
Wilmington High School

Site Coordinator

Mary Anne Wells
Mathematics and Science Resource Center
University of Delaware, Wilmington

Jillann Hounsell, Teacher
Henry B. du Pont Middle School, Hockessin

Thomas Hounsell, Teacher
Henry B. du Pont Middle School, Hockessin

Martin J. Cresci, Teacher
Henry C. Conrad Middle School, Wilmington

Maryland

Secondary Science Coordinator
Montgomery County Public Schools
Gerry Consuegra

Site Coordinator

Patricia Hagan, Science Project Specialist
Montgomery County Public Schools

Vince Parada, Teacher
Rosa M. Parks Middle School, Olney

Pam Fountain, Teacher
Tilden Middle School, Rockville

Oregon

Site Coordinator

Angie Ruzicka
4J Schools—Eugene School District

Courtney Abbott, Teacher
Kelly Middle School, Eugene

Julie Hohenemser, Teacher
Cal Young Middle School, Eugene

Judy Francis, Teacher
Roosevelt Middle School, Eugene

Pennsylvania

Site Coordinator

James L. Smoyer
Boyce Middle School, Pittsburgh

Sherri Petrella, Teacher
David E. Williams Middle School, Coraopolis

Jennifer Robinson
David E. Williams Middle School, Coraopolis

Jean M. Austin
Fort Couch Middle School, Upper St. Clair

Tennessee

Site Coordinator

Jimmie Lou Lee
Center for Excellence for Research
and Policy on Basic Skills
Tennessee State University, Nashville

Janet Zanetis, Teacher
Ellis Middle School, Hendersonville

Judy W. Laney, Teacher
Grassland Middle School, Franklin

Gary Mullican, Teacher
Central Middle School, Murfreesboro

Ann Orman, Teacher
West End Middle School, Nashville

The NSRC appreciates the contribution of its
STC/MS project evaluation consultants—

Program Evaluation Research Group (PERG), Lesley College

Sabra Lee
Researcher, PERG

George Hein
Director (retired), PERG

Center for the Study of Testing, Evaluation,
and Education Policy (CSTEEP), Boston College

Joseph Pedulla
Director, CSTEEP

Maryellen Harmon
Director (retired), CSTEEP

Preface

Community leaders and state and local school officials across the country are recognizing the need to implement science education programs consistent with the National Science Education Standards to attain the important national goal of scientific literacy for all students in the 21st century. The Standards present a bold vision of science education. They identify what students at various levels should know and be able to do. They also emphasize the importance of transforming the science curriculum to enable students to engage actively in scientific inquiry as a way to develop conceptual understanding as well as problem-solving skills.

The development of effective standards-based, inquiry-centered curriculum materials is a key step in achieving scientific literacy. The National Science Resources Center (NSRC) has responded to this challenge through Science and Technology Concepts–Secondary™. Prior to the development of these materials, there were very few science curriculum resources for secondary students that embodied scientific inquiry and hands-on learning. With the publication of STC–Secondary™, schools will have a rich set of curriculum resources to fill this need.

Since its founding in 1985, the NSRC has made many significant contributions to the goal of achieving scientific literacy for all students. In addition to developing Science and Technology Concepts–Elementary™—an inquiry-centered science curriculum for grades K through 6—the NSRC has been active in disseminating information on science teaching resources, preparing school district leaders to spearhead science education reform, and providing technical assistance to school districts. These programs have had a significant impact on science education throughout the country. The transformation of science education is a challenging task that will continue to require the kind of strategic thinking and insistence on excellence that the NSRC has demonstrated in all of its curriculum development and outreach programs. The Smithsonian Institution, our sponsoring organization, takes great pride in the publication of this exciting new science program for secondary students.

Letter to the Students

 Smithsonian Institution
National Science Resources Center

Dear Student,

The National Science Resources Center's (NSRC) mission is to improve the learning and teaching of science for K-12 students. As an organization of the Smithsonian Institution, the NSRC is dedicated to the establishment of effective science programs for all students. To contribute to that goal, the NSRC has developed and published two comprehensive, research-based science curriculum programs: Science and Technology Concepts-Elementary™ and Science and Technology Concepts-Secondary™.

By using the STC-Secondary™ curriculum materials, we know that you will build an understanding of important concepts in life, earth, and physical sciences; learn critical-thinking skills; and develop positive attitudes toward science and technology. The National Science Education Standards state that all secondary students "...should be provided opportunities to engage in full and partial inquiries.... With an appropriate curriculum and adequate instruction, ... students can develop the skills of investigation and the understanding that scientific inquiry is guided by knowledge, observations, ideas, and questions."

STC-Secondary also addresses the national technology standards published by the International Technology Education Association. Informed by research and guided by standards, the design of the STC-Secondary units address four critical goals:

- Use of effective student and teacher assessment strategies to improve learning and teaching

- Integration of literacy into the learning of science by giving students the lens of language to focus and clarify their thinking and activities

- Enhanced learning using new technologies to help students visualize processes and relationships that are normally invisible or difficult to understand

- Incorporation of strategies to actively engage parents to support the learning process

We hope that by using the STC-Secondary curriculum you will expand your interest, curiosity, and understanding about the world around you. We welcome comments from students and teachers about their experiences with the STC-Secondary program materials.

Sally Goetz Shuler
Executive Director
National Science Resources Center

Navigating an STC–Secondary™ Student Guide

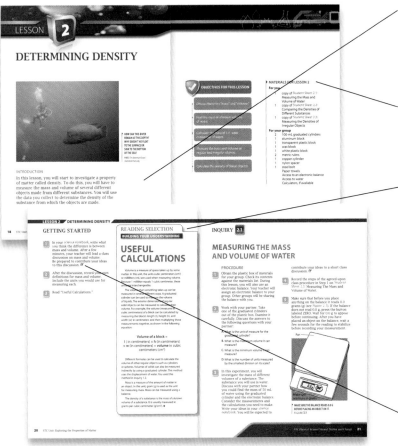

INTRODUCTION
This short paragraph helps get you interested about the upcoming inquiries.

MATERIALS
This helps you get organized and prepare for your inquiries.

READING SELECTION:
BUILDING YOUR UNDERSTANDING
These reading selections are part of the lesson, and give you information about the topic or concept you are exploring.

NOTEBOOK ICON 📝
During the course of an inquiry, you'll record data in different ways. This icon lets you know to record in your science notebook. Student sheets are called out when you're to write there. You may go back and forth between your notebook and a student sheet. Watch carefully for the icon throughout the procedure.

SAFETY TIPS
Safety in the science classroom is very important. Tips throughout the student guide will help you to practice safe techniques while conducting investigations. It is very important to read and follow all safety tips.

SAFETY TIP

PROCEDURE

This tells you what to do. Sometimes the steps are very specific, and sometimes they guide you to come up with your own investigation and ways to record data.

REFLECTING ON WHAT YOU'VE DONE

These questions help you think about what you've learned during the lesson's inquiries, apply them to different situations, and generate new questions. Often you'll discuss your ideas with the class.

READING SELECTION: EXTENDING YOUR KNOWLEDGE

These reading selections come after the lesson, and show new ways that the topic or concept you learned about during the lesson can be applied, often in real-world situations.

GLOSSARY

Here you can find scientific terms defined.

INDEX

Locate specific information within the student guide using the index.

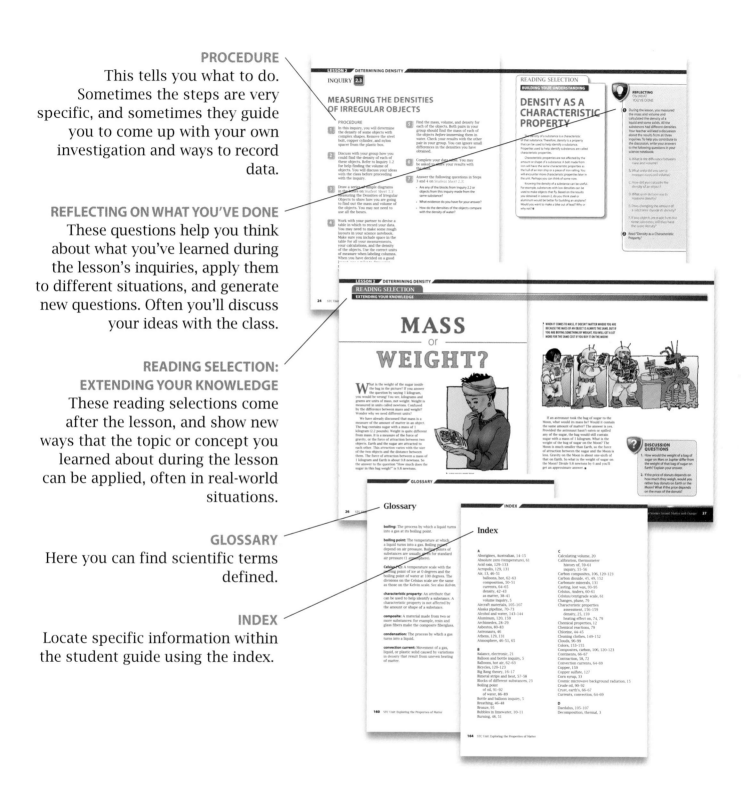

Contents

CONTENTS

THE NATURE OF MATTER

▸ SOME MATTER, SUCH AS THIS BISMUTH, FORM CRYSTALS. IS THIS BISMUTH A PURE SUBSTANCE OR A MIXTURE OF SUBSTANCES?

PHOTO: T.K. Ives, Jr.

INTRODUCTION

What is the nature of matter? What are its parts, or components? How do these components combine, break down, and react with each other? These are some of the questions you will answer in this unit. In this lesson, you will consider what you already know about matter. You may know something about its phases (or states), properties, and reactions—and its components. You will complete a circuit of eight inquiries on pure substances and mixtures. These inquiries are designed to get you thinking about what makes up matter and how it behaves. The observations you make and ideas you discuss in this lesson will play a key role in the inquiries that you will perform later in this unit.

OBJECTIVES FOR THIS LESSON

Discuss your understanding of the terms pure substance and mixture.

Observe matter in the form of pure substances and mixtures.

Use your own words and ideas to explain your observations.

▶ MATERIALS FOR LESSON 1

For you

1   copy of Student Sheet 1: Our Ideas About Pure Substances and Mixtures

1   pair of safety goggles

GETTING STARTED

 Your teacher will place you in groups of four. What does your group know about the nature of matter? Discuss with your group what you know about the measurement of matter, its states, its characteristic properties, and its reactions, and record it on Student Sheet 1: Our Ideas About Pure Substances and Mixtures.

 Share your ideas in the class discussion, and record your classmates' ideas.

 With your group, discuss the differences between a pure substance and a mixture. How do you know if something is a mixture or a pure substance? Provide some examples of mixtures, and some examples of pure substances.

▶ **WHAT DO YOU ALREADY KNOW ABOUT DIFFERENT KINDS OF SUBSTANCES?**

PHOTO: Courtesy of Bristol-Myers Squibb/Charlotte Raymond, Photographer

4 Share your ideas and examples in a class discussion.

5 In this lesson, you will investigate the properties and behaviors of a variety of mixtures and pure substances. Working with another student, you will complete eight inquiries. You will be graded on the seriousness of your efforts, on the carefulness of your observations, and on your cooperation with your lab partner.

6 Listen carefully to the safety instructions given by your teacher. In order to participate in the lab work for this class, you are expected to follow safe laboratory procedures.

7 Each inquiry station has a basic topic. Each pair of students will start at a different inquiry station. At each station you will follow the instructions on the Inquiry Card. These instructions are also in your student guide. When you make observations or think you can explain what you are observing, you should discuss these ideas with your partner. Remember: Exchanging ideas with others is a very important part of science. You will have 4-5 minutes to complete each inquiry and record your observations on Student Sheet 1: Our Ideas About Pure Substances and Mixtures.

8 Your teacher will set up two sets of identical inquiries—A and B. You will be assigned to either A or B. Place an asterisk beside your first inquiry on the student sheet so that you write your answers in the correct spaces.

9 When you have completed each inquiry, put the apparatus back as you found it at the beginning of the experiment.

10 If you have any questions about the procedure, you should ask your teacher now.

SAFETY TIPS

Wear safety goggles at all times during the inquiries.

Keep long hair tied back.

Be careful with hydrochloric acid. If acid gets on your clothes or skin, wash it off immediately with lots of water.

Tell your teacher immediately about any accidents involving acid.

INQUIRY **1.1**

FINDING THE CONDUCTOR

PROCEDURE

1 Set up a circuit with the batteries in holders, a lightbulb, a bulb holder, and the connector wires (see Figure 1.1). Touch the unfastened alligator clips together and observe. Answer the following question on Student Sheet 1: What happens to the bulb when you complete the circuit by touching the alligator clips?

2 Place the copper cylinder in the circuit between the alligator clips, and observe the bulb. Answer the following question: What happens to the lightbulb when the copper cylinder is placed in the circuit?

3 Connect the ends of the small pencil with the alligator clips, and observe the bulb. What happens to the bulb with the pencil in the circuit? Record your observations and ideas.

4 Finally, use the connector wires to put the piece of zinc in the circuit, and observe the bulb. Answer the following question: What happens to the bulb with the zinc in the circuit?

5 What conclusions can you make about the ability of these substances to conduct electricity? Record your ideas.

6 Disconnect the circuit, and return the apparatus to its original condition for the next group.

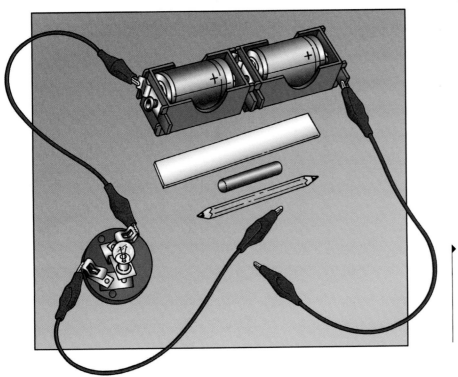

► PLACE EACH OF THE OBJECTS TO BE TESTED IN THE CIRCUIT, ONE AT A TIME.
FIGURE **1.1**

FILTERING A MIXTURE

PROCEDURE

1 Observe the mixture in the jar labeled Mixture A. Does it appear to be more than one substance? Describe it on Student Sheet 1.

2 Place a lab scoop of the mixture in your test tube and add 10 mL of water.

3 Place a stopper on the test tube, and shake the test tube for about 30 seconds. (Be careful not to hit the test tube against your lab table or another hard object.)

4 What happens to the contents of the test tube? Record your observations.

5 Fold a filter paper (see Figure 1.2) and place it in the funnel. Pour the contents of the test tube through the funnel into the second test tube.

6 Describe the appearance of the substance on the filter paper.

7 What do you think happened to the parts of the original mixture once you added the water and filtered it?

8 How do you think you might get back all the parts of the original mixture?

9 Rinse out the test tubes and funnel and throw away the used filter paper.

10 Replace the apparatus for the next group.

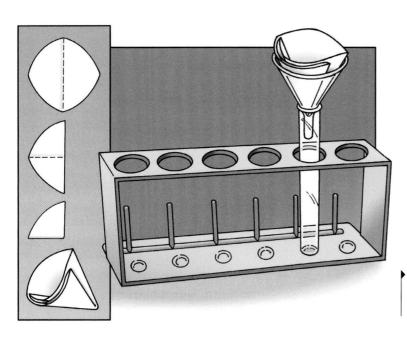

▸ FOLD THE FILTER PAPER TWICE IN TWO DIFFERENT DIRECTIONS AS SHOWN.
FIGURE **1.2**

INQUIRY **1.3**

THE BURNING CANDLE

PROCEDURE

1 Use a match to light the candle.

2 What can you see taking place at or near the top of the candle? Write your observations on Student Sheet 1.

3 Place the open end of the beaker over the candle (see Figure 1.3). Let the beaker stay over the candle for a few minutes.

4 What happened after the beaker was placed over the candle? Record your observations.

5 Why do you think the candle reacted the way it did? Record your answer.

6 Restore the apparatus to its original condition for the next group.

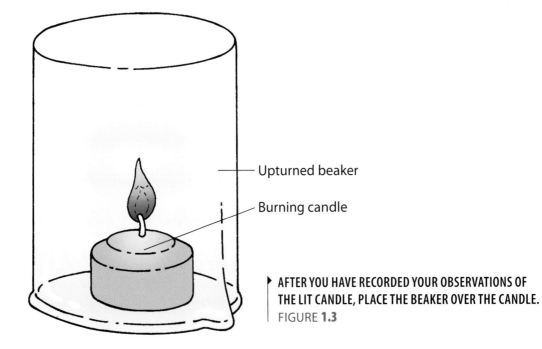

— Upturned beaker

— Burning candle

▸ AFTER YOU HAVE RECORDED YOUR OBSERVATIONS OF THE LIT CANDLE, PLACE THE BEAKER OVER THE CANDLE.
FIGURE **1.3**

SEPARATING THE MIXTURE

ADDING THE ACID

PROCEDURE

1 Place a lab scoop of Substance B from the jar in the petri dish.

A. Use your loupe to observe the mixture.

B. Describe the appearance of the mixture on Student Sheet 1.

C. Take the magnet and move it across the bottom of the petri dish (see Figure 1.4).

D. What do you observe happening to the mixture now?

E. What could account for your observations?

F. Return the substance to the jar, and set up the apparatus for the next group.

PROCEDURE

1 Set up the apparatus as shown in Figure 1.5.

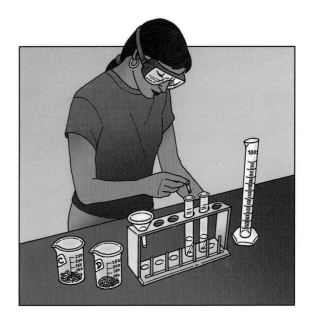

▶ SET UP FOR INQUIRY 1.5
FIGURE **1.5**

2 Carefully pour 10 mL of dilute hydrochloric acid through the funnel into each of two test tubes. Note: If you spill the acid on yourself or your clothes, wash it off immediately with lots of cold water and notify your teacher.

▶ PLACE THE MAGNET UNDERNEATH THE PETRI DISH OF THE MIXTURE.
FIGURE **1.4**

Inquiry 1.5 continued

INQUIRY  **1.6**

3 Add one piece of Substance C to the first test tube. What do you observe happening? (Touch the outside of the test tube to detect any temperature changes. Listen for any reaction in the test tube.) Describe what happens on Student Sheet 1.

4 Add one piece of Substance D to the second test tube and observe what happens. Describe the reaction in the test tube. Did you feel any temperature change? Did you hear anything?

5 How would you compare the reactions of the two substances with the hydrochloric acid?

6 Dispose of the hydrochloric acid as directed by your teacher, and rinse out the test tubes. Return the equipment to the way you found it for the next group.

COMPARING THE TWO MIXTURES

PROCEDURE

1 Place a lab scoop of Substance E in the first test tube and add 10 mL of water.

2 Describe the appearance of the mixture formed in the first test tube on Student Sheet 1. Use your loupe to observe it more closely.

3 Place a lab scoop of Substance F in the second test tube, and add 10 mL of water.

4 How does the mixture in the second test tube appear? Record your observations.

5 How might you account for the differences in the two substances when water was added to them?

6 Empty out the contents of the two test tubes, rinse out the test tubes, and set them up for the next group.

REACTING A TABLET

PROCEDURE

1 Add 20 mL of water to the test tube.

2 Put the thermometer in the test tube (see Figure 1.6).

3 Wait 30 seconds and record the temperature on Student Sheet 1.

4 Drop one piece of the white tablet into the water.

5 For the next 3 minutes, carefully observe what happens.

6 What happened after the tablet was added to the water? Record your observations.

7 When no further changes take place, record the temperature of the water.

8 When you finish, empty and rinse the test tube for the next group.

▶ MAKE SURE YOU WAIT 30 SECONDS AFTER PLACING THE THERMOMETER IN THE TEST TUBE BEFORE YOU READ THE TEMPERATURE.

FIGURE **1.6**

 INQUIRY **1.8**

MIXING THE SOLUTIONS

PROCEDURE

1 Use the graduated cylinder to measure 10 mL of Solution G. Pour it into a test tube. Describe the appearance of the solution in the test tube and write your observations on Student Sheet 1. Set the test tube in the test tube rack.

 Place an equal amount of Solution H in the second test tube. Do not use the graduated cylinder; it is okay to approximate the amount. Describe the appearance of the solution.

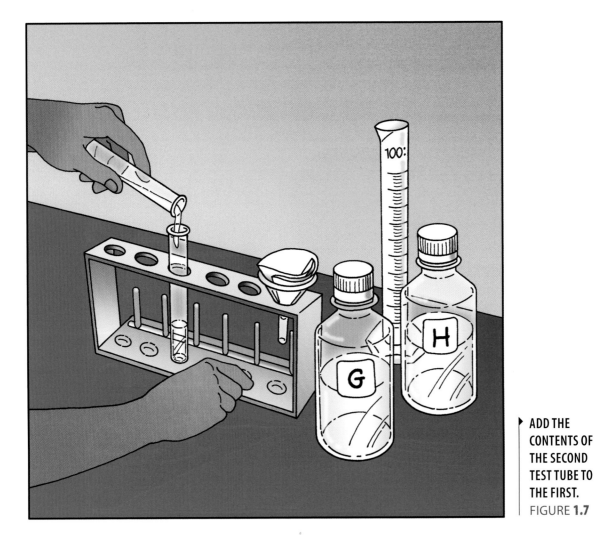

▶ ADD THE CONTENTS OF THE SECOND TEST TUBE TO THE FIRST.

FIGURE **1.7**

3 Add the contents of the second tube to the first test tube (see Figure 1.7). Describe the appearance of the resulting substance.

4 What do you think happened when you mixed the two solutions? What do you think would happen if you filtered the contents of the test tube? Try it. Record your observations.

5 Throw away the used filter paper. Empty the contents of the test tube in the place designated by your teacher. Wash both test tubes and the funnel in the soapy water supplied by your teacher. Rinse and return them to the test tube rack for the next group of students.

REFLECTING
ON WHAT YOU'VE DONE

1 Return to your group. You will receive a sheet of newsprint with the title of an inquiry on it.

A. Brainstorm with your group, and come up with one observation you made or one conclusion you reached from the inquiry. Record it on the sheet. You will have 1 minute.

B. After 1 minute, move to the next sheet of newsprint. Your group will visit a newsprint sheet for each inquiry in the circuit and record an observation or conclusion for each.

2 Participate in a discussion about the different observations the groups made.

PURE SUBSTANCE OR MIXTURE?

INTRODUCTION

The characteristics of substances, such as density, and the behavior of substances when they are heated can be used to help identify substances. However, there is one problem. These properties are most useful in identifying pure substances. Many of the materials that we come across in our daily lives are not pure. Mixtures of substances are much more common than pure substances. For example, look at your own body. You are made up of matter that consists of many complex substances that work together to produce the chemical reactions that occur in living organisms.

Identifying the individual substances from which living things are made is very difficult. To separate the substances in a living cell, a biochemist would need to grind up samples of the tissue and then expose the souplike mixture to an array of separation techniques to obtain pure samples of each substance.

Finding out whether something is pure is hard work! In this lesson, you will try to define the term "pure substance." You will devise your own techniques to determine whether eight different samples are pure substances or mixtures. You will then discuss the difficulties you encountered in classifying the samples.

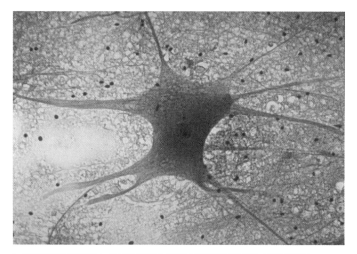

▶ THIS NERVE CELL, LIKE EVERY CELL IN YOUR BODY, CONTAINS THOUSANDS OF DIFFERENT SUBSTANCES. EACH ONE PERFORMS A DIFFERENT FUNCTION WITHIN THE CELL.

PHOTO: © Carolina Biological Supply Company, used with permission

OBJECTIVES FOR THIS LESSON

▸ Discuss the meaning of the term "pure substance."

▸ Discuss how you can distinguish between pure substances and mixtures.

▸ Use your own techniques to discover whether several samples of matter are pure substances or mixtures.

▸ MATERIALS FOR LESSON 2

For you

1   copy of Student Sheet 2.1: Identifying Pure Substances and Mixtures

1   pair of safety goggles

For your group

8   samples (labeled A through H)

2   loupes (double-eye magnifiers)

2   lab scoops

2   pipettes

4   petri dish lids or bases

4   sheets of black paper

4   sheets of white paper

1   magnet

4   test tubes

1   test tube rack

1   test tube brush

1   plastic cup

1   label

    Access to water

GETTING STARTED

INQUIRY  2.1

1 Before you start investigating whether substances are pure or mixtures, it would be useful to think about how you already use these terms. Write your definition of a pure substance on Student Sheet 2.1: Identifying Pure Substances and Mixtures. Give two examples of pure substances. For each, explain why you think it is pure. If you were given a sample of unknown matter, how could you tell whether it was a pure substance or a mixture? Record your ideas.

2 Use your answers to contribute to a class discussion and concept map.

SAFETY TIP

Wear your safety goggles throughout the inquiry.

Do not taste any of the substances.

▶ MILK LOOKS LIKE A SINGLE SUBSTANCE. IS IT PURE, OR IS IT A MIXTURE?

PHOTO: Stephen Ausmus, Agricultural Research Service/U.S. Department of Agriculture

DETERMINING WHETHER SUBSTANCES ARE PURE OR MIXTURES

PROCEDURE

1 Have one member of your group obtain the plastic box of materials. Check its contents against the list of materials.

2 Take samples A through H out of the plastic box.

3 The purpose of this inquiry is to answer the question, "Which of these substances are mixtures and which are pure substances?" You have about 20 minutes to answer this question and record your answers, so you will need to divide the work among the members of your group.

4 You may use all of the apparatus in the plastic box, plus water, to help you with your investigation. Devise your own techniques to determine whether each sample is a pure substance or a mixture.

5 For each sample, record your findings in Table 1 on Student Sheet 2.1: Identifying Pure Substances and Mixtures.

6 Use any additional data collected by other group members to complete Table 1. Discuss the results with the other members of your group.

7 Record your answers to the following questions:

- How can the properties of pure substances be used to discover whether a sample is a mixture?

- Were the samples all well-mixed? How did the extent of mixing affect your investigation?

8 Put any waste left over from each sample in the appropriate container. Wash all of the test tubes and return the materials to the plastic box. Make sure you also wash your hands when you are finished.

9 Your teacher will lead a class discussion about your procedures and results. Listen carefully as other students describe their approaches to answering the question of whether a sample is a pure substance or a mixture, explain their results, and make suggestions for alternative approaches to the problem.

REFLECTING
ON WHAT
YOU'VE DONE

1 Look at the concept map the class created during "Getting Started." Have any of your ideas changed?

2 Revisit your definition of "pure substance" from the start of the lesson. If it is different from the one agreed on in class, write the new one on Student Sheet 2.1.

3 Your teacher will show you a blue solution. How would you describe the appearance of the solution? Write your description in your science notebook. Think about the questions your teacher raises about the solution.

4 Label your group's cup of solution. You will begin the next lesson by examining your group's cup of blue solution.

PERFECT TEAMWORK

▶ WHEN BUILDING THE STEALTH BOMBER, DESIGNERS USED MANY COMPOSITE MATERIALS IN PLACE OF METAL. THE USE OF THESE MATERIALS, COMBINED WITH ITS SPECIAL SHAPE, MAKES THE PLANE INVISIBLE TO MOST RADAR.

PHOTO: U.S. Air Force photo/Staff Sgt. Bennie J. Davis III

Wouldn't it be great to have a baseball team made entirely of the world's greatest pitchers? Well, no. This would not be a happy team. It wouldn't matter how many strikes these superstars could throw. A team without players who are good at catching, hitting, and stealing bases would have a hard time winning. A team with a good balance of skills is more likely to make it to the World Series.

Combining skills is important for making strong materials, too. Often, a pure substance on its own does not have all of the necessary properties for a particular material. But you can make many useful materials by combining two or more substances that have different properties. The result is a mixture called a composite. A good composite exploits the best properties of each ingredient.

▶ WHY ARE FIBERGLASS AND CARBON FIBER COMPOSITES GOOD MATERIALS FOR MAKING FISHING RODS?

PHOTO: Bugeater/ creativecommons.org

People have been making composites since the beginning of civilization. For ancient peoples, dried mud and even animal dung were handy for making huts. The huts were simple to make: Find dirt, add water. The mud kept out the wind and didn't rot, but it crumbled and cracked. Ancient peoples also used straw, grass, or sticks, which were woven into durable mats, to make hut walls. But woven walls leaked.

The solution was to combine the two. In many parts of the world, people realized they could weave a house frame (usually supported by timber) out of straw, grass, or sticks and cover it with mud. The result, called "wattle-and-daub" construction, wasn't always pretty. But it kept out the cold and did not fall apart every time the kids got a little rowdy.

▶ THE MASAI OF KENYA HAVE BEEN USING COMPOSITE MATERIALS—MUD, DUNG, AND STRAW— TO MAKE HOMES FOR THOUSANDS OF YEARS.

PHOTO: © David Marsland

READING SELECTION

EXTENDING YOUR KNOWLEDGE

▶ IN THE EARLY 1900S, SOLID WOODEN RACKETS WERE USED TO PLAY TENNIS.

PHOTO: Library of Congress, Prints & Photographs Division, LC-B2-2278-7

▶ MODERN RACKETS LIKE THIS ONE, MADE FROM SEVERAL DIFFERENT TYPES OF COMPOSITES, ARE MUCH STRONGER AND LIGHTER THAN WOODEN RACKETS.

PHOTO: National Science Resources Center

People have been coming up with new composite materials ever since. Usually, a composite has two materials with opposite properties. The two materials put together as a composite can do what each ingredient alone cannot. Like ancient wattle-and-daub huts, modern composites are often made of fibers embedded in a solid that sticks the fibers together. The fibers are strong but floppy. The solid isn't floppy, but it easily shatters or cracks. A combination of these two opposites can be unbeatable. The solid provides stiffness. The fibers keep the solid from cracking apart. (This is because a crack would have to break too many of the strong fibers running through the solid.)

Fiberglass is one example of a modern composite. To make fiberglass, glass is melted and stretched into long threads. The glass threads are woven into cloth. The cloth is embedded in plastic goo, and the whole thing is shaped in a mold. When the goo hardens, the object has the shape of the mold, and is lightweight and cheap to make. Fiberglass was originally developed to cover radar dishes on World War II bombers. It is now used for everything from boats to fishing rods to picnic tables.

More recently, engineers have developed new composite materials. One of these composites contains carbon fibers that are stiffer and much more heat-resistant than glass. A given weight of carbon composite is stronger than steel. This lightweight strength makes carbon composites ideal for use in many types of objects that would normally be made of metal. The wings of jet fighter planes and helicopter blades are two examples.

Composites are widely used in sports equipment and are replacing many natural materials. For example, tennis rackets, originally made from wood, now have frames made from glass, carbon, or boron fibers embedded in a plastic-like nylon. The core of the racket is made from a plastic foam. The result is a lightweight, stiff racket that is easy to control and that returns the ball with maximum force.

Even though carbon composites are one of the latest advances in composite materials, they share something with ancient mud huts. Both combine the best parts of different materials to make something that is better than either one alone. ■

DISCUSSION QUESTIONS

1. What is one object in your home that is made from a composite material? What is the function of that object, and why was that composite chosen for that function?

2. Adobe is a building material that is found in the oldest buildings on earth. Use library or Internet resources to study the composition and history of adobe.

SEPARATING A SOLUBLE AND AN INSOLUBLE SUBSTANCE

INTRODUCTION

Sewage is very smelly stuff. It is also an interesting mixture. It usually looks like a cloudy brown liquid with lumps floating in it. It is difficult to believe that after it is processed in a sewage facility, it can be released as a clear liquid into a river. Cleaning sewage is a complex process. Workers in treatment facilities apply scientific knowledge of the different properties of the substances found in sewage to remove them from the water. Many types of separation techniques and treatments are involved. Some of the large pieces are removed by a screen. Some of the smaller particles are removed by allowing them to settle out. Other particles are filtered out. Chemicals are added to speed up the cleaning process, and microorganisms are cultivated to eat some of the sewage. Separation techniques play a big part in keeping our rivers clean, and a big part in our lives in general. In this lesson, you will apply your knowledge of solutions, separation techniques, and the phases of matter to a separation problem less smelly than the separation of sewage!

▶ SEWAGE IS A MIXTURE. WHY IS KNOWLEDGE OF SEPARATION TECHNIQUES IMPORTANT TO ENSURE THE HEALTH OF OUR RIVERS?

PHOTO: David Weekly/creativecommons.org

OBJECTIVES FOR THIS LESSON

Discuss evaporation as a separation technique.

Filter mixtures containing water.

Design and conduct an inquiry to clean rock salt.

MATERIALS FOR LESSON 3

For you

1   copy of Student Sheet 3.1: Filtering a Solution

1   copy of Student Sheet 3.2: Cleaning Rock Salt

1   pair of safety goggles

For your group

     Plastic cup of blue solution from Lesson 2 (labeled with the names of your group)

1   plastic cup

1   jar containing rock salt

1   jar containing zinc oxide

2   funnels

6   filter papers

2   100-mL graduated cylinders

2   lab scoops

2   loupes (double-eye magnifiers)

4   test tubes

2   test tube racks

2   250-mL beakers

2   plastic spoons

GETTING STARTED

1. One member of your group should obtain the plastic box containing the materials. Another student should get your group's plastic cup from Lesson 2.

2. Use a magnifying loupe to examine the contents of the plastic cup.

3. Discuss the following questions with the other members of your group:

   A. What do you think this blue substance is?

   B. How did it get there?

   C. Where did the water go?

4. Your teacher will ask you about your ideas and observations and will discuss the processes involved in forming the blue substance.

5. Predict what will happen to the substance if you add 25 mL of water to the cup.

6. Test your prediction by adding 25 mL of water to the cup. Use the plastic spoon to stir the contents.

7. Think about the answers to the following questions:

   A. What happens after you add 25 mL of water to the blue substance?

   B. What are the properties of the mixture?

   C. What do you think you have made?

   Be prepared to participate in a class discussion.

8. Divide your solution into two approximately equal parts. For the remainder of the lesson, you will work with your lab partner. Divide the apparatus in the plastic box between the two pairs in your group.

SAFETY TIP

Wear your safety goggles throughout the lesson.

▶ **WHAT DO YOU THINK WILL HAPPEN WHEN YOU ADD WATER TO THIS SUBSTANCE?**

PHOTO: National Science Resources Center

FILTERING A SOLUTION

PROCEDURE

1 Your teacher will show you how to fold a piece of filter paper and insert it into a funnel. As shown in this demonstration and in Figure 3.1, fold the paper and fit it inside the funnel. Wet the paper with a few drops of water so that it sticks to the funnel walls. Observe whether the water passes through the filter paper.

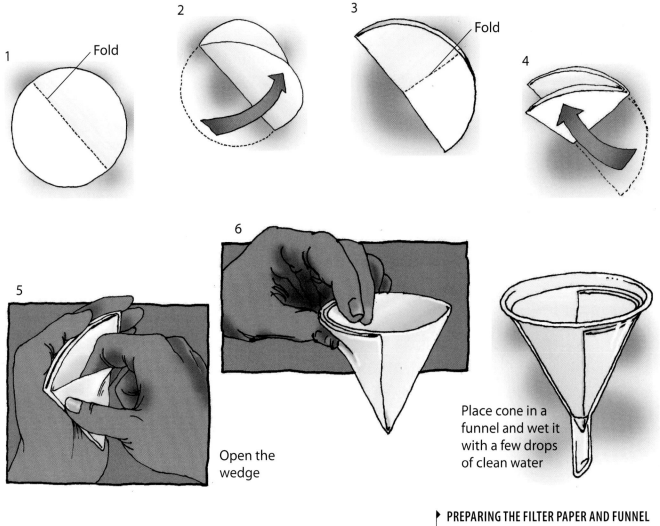

1
Fold

2

3
Fold

4

5

6
Open the wedge

Place cone in a funnel and wet it with a few drops of clean water

▶ **PREPARING THE FILTER PAPER AND FUNNEL**
FIGURE **3.1**

Inquiry 3.1 continued

2 What do you think will happen to the copper (II) sulfate solution if you pour it into the funnel? Record your prediction in Table 1 on Student Sheet 3.1: Filtering a Solution.

3 Place a test tube in the test tube rack. Place the funnel with the filter paper into the test tube.

4 Test your prediction by pouring the solution into the funnel (see Figure 3.2). Make sure the solution does not go over the edge of the paper. Record your result in Table 1.

5 Dispose of the filter paper. Fold a new one and place it in the funnel.

6 Add one lab scoop of zinc oxide to approximately 10 mL of water in a 250-mL beaker. Stir the mixture with a plastic spoon. What will happen when you filter this mixture? Record your prediction in Table 1.

7 Use a clean test tube to repeat the filtration procedure. Record your result in Table 1.

8 Discuss your observations with your partner and answer the following questions. Be prepared to participate in a class discussion of your results and to explain your ideas.

- What effect did filtration have on the two mixtures?

- Can you think of a property of solutions other than those you discovered?

9 Use the waste containers provided to dispose of the copper (II) sulfate solution and the zinc oxide solution. Place your used filter papers in the trash can. Rinse the apparatus and return it to the plastic box.

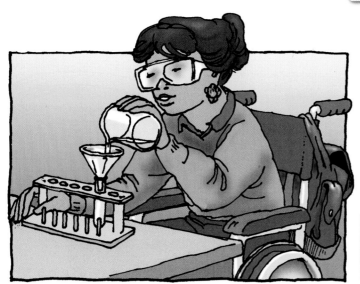

▶ AFTER SETTING UP YOUR APPARATUS, POUR THE COPPER (II) SULFATE SOLUTION INTO THE FUNNEL CONTAINING THE FILTER PAPER.
FIGURE **3.2**

INQUIRY 3.2

CLEANING ROCK SALT

PROCEDURE

1. Put four lab scoops of rock salt into the plastic cup. Examine it with the magnifying loupe. Write a description of the rock salt on Student Sheet 3.2: Cleaning Rock Salt.

2. Most of the salt used in food is made from rock salt. Discuss these questions with your partner:

   A. Would you want to eat this sample?

   B. Do you think it is pure?

   C. What do you think the contaminants could be?

3. How could you use the remaining apparatus you have been given to obtain only the soluble component of the rock salt? Record your answers to the following questions:

   • What are you trying to do?

   • What materials will you use?

4. Record the procedure you and your partner devised.

5. Check your ideas with your teacher.

6. Follow your procedure to purify the salt. If you have any problems, consult your teacher.

▶ YOU EAT THIS ROCK. WHAT IS IT AND HOW IS IT PURIFIED?

PHOTO: NASA African Monsoon Multidisciplinary Analyses (NAMMA)

READING SELECTION

SEPARATING SOLIDS FROM LIQUIDS

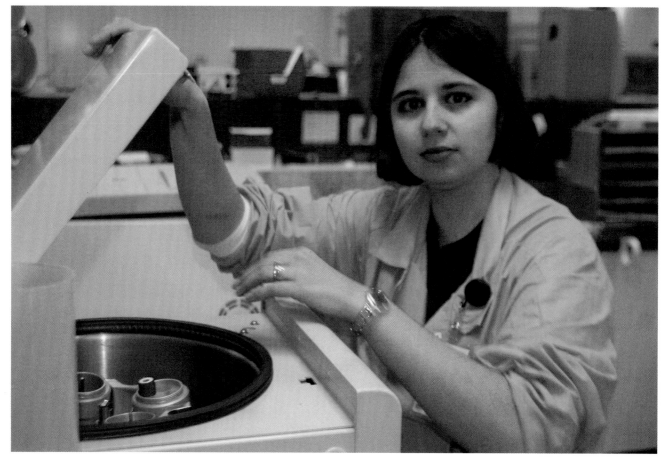

▶ A CENTRIFUGE IS A MACHINE THAT CAN BE USED TO SEPARATE INSOLUBLE SOLIDS FROM LIQUIDS.

PHOTO: Ellsworth Air Force Base photo by Airman Nathan Riley

▶ THESE POTASSIUM FERRICYANIDE CRYSTALS WERE PRODUCED WHEN THE WATER IN A SATURATED SOLUTION OF POTASSIUM FERRICYANIDE WAS SLOWLY EVAPORATED.

PHOTO: Wikimedia Commons

When separating substances, it is important to choose the correct separation technique, or method, to separate the components of a mixture from one another. For example, insoluble solids can be separated from liquids in several different ways. The technique used depends on how well substances are mixed together.

To separate insoluble impurities from salt, you used a process called filtration. The filter paper allowed the soluble solute (salt) and the solvent (water) to pass through, but it trapped the larger pieces of insoluble impurities as a residue. The substances that pass through the filter paper are called the filtrate.

Large pieces of insoluble substances will often settle out of a mixture of a solid and a liquid. This process is called sedimentation, because the solid forms sediment on the bottom of the container. If the solid is very fine, this process can be speeded up with a machine called a centrifuge. Centrifuges spin a test tube very fast, and the solid moves quickly to the bottom of the test tube.

To separate a solid solute from a solvent (like salt from water), you used evaporation. At room temperature, water evaporates from a solution very slowly. But the rate of evaporation can be accelerated by heating the solution. As the water evaporated from your salt and water solution, the solution became more concentrated. Eventually, a saturated solution of salt formed. As more water evaporated, the salt crystallized into white crystals. When crystallization happens slowly, big crystals form. Small crystals form when crystallization happens quickly. Crystalline solids have unique crystal shapes. Therefore, crystal shape is a characteristic property of a substance. ∎

REFLECTING
ON WHAT YOU'VE DONE

1. As a class, discuss the procedures used by the different pairs.

2. Read "Separating Solids From Liquids."

3. Your cleaned salt sample will not be obtained until a later lesson. When you get a sample of clean solid salt, look at it closely and answer the following questions:

A. Are any crystals present?

B. Are they all the same shape?

C. How clean is your salt?

D. Is there any evidence that it is still not pure?

E. If your salt is still not pure, can you suggest why?

Separating Solutions and the Salty Sea

▶ THIS CARAVAN OF BARGES IS CARRYING SALT FROM PRODUCTION FACILITIES IN MEXICO.

PHOTO: pearlbear/creativecommons.org

Why is the sea salty? Where does all that salt come from? How does it get there? Much of the salt comes from the land. When it rains, rainwater dissolves soluble substances, including common salt (sodium chloride), from soil and rocks. Some of these substances eventually find their way into creeks and rivers, and from there, they are carried to the sea.

Why is the sea saltier than rivers? Once in the sea, soluble substances are concentrated. Heat from the sun evaporates the water from the sea but leaves the salts behind. Over millions of years, seas become saltier and saltier. For the same reason, lakes that have no outlet to the sea quickly become salty. Lakes can even be saltier than seas.

Salt is a valuable commodity that has been traded for thousands of years. It is used in food, providing flavor and acting as a food preservative. It is also used as a starting material to make a wide variety of other chemicals. These chemicals are used in many industrial processes, including making glass, soap, and chlorine.

▶ SALT MAKING BY EVAPORATION, SALT LAKE, UTAH (EARLY 20TH CENTURY)

PHOTO: Library of Congress, Prints & Photographs Division, LC-USZ62-70599

Today, most salt comes from mines, although a lot is also extracted from the sea or salty lake water. Salt has been extracted from salty bodies of water throughout history. One common method of extraction is to let the heat from the sun completely evaporate seawater that is trapped in pools or small lagoons.

In some desert areas, water is very scarce. But many of these deserts are near seas (or salt lakes), and salt can be removed from seawater to get fresh water. This process is called desalination. Seawater that is desalinated is fresh enough to drink and grow crops. In some desalination plants, distillation is used to separate the salt out from the water. First, the saltwater is heated. The water evaporates away from the salt as steam. Then the steam condenses to form fresh water. This process requires a lot of energy, so it is very expensive. Other desalination plants remove salt from water by a process called reverse osmosis. ■

DISCUSSION QUESTIONS

1. How is evaporation important in salt mining?

2. Use library or Internet resources to research the Salton and Aral Seas. What do they have in common? What environmental problems do they have?

▶ THE YUMA DESALTING PLANT IN ARIZONA IS ONE OF THE WORLD'S LARGEST DESALINATION PLANTS, CAPABLE OF PRODUCING 72 MILLION GALLONS OF DESALTED WATER PER DAY.

PHOTO: Andy Pernick/U.S. Bureau of Reclamation

SEPARATING SOLUTES

▶ IS THIS MONEY
GENUINE? WOULD YOU
BE ABLE TO FIND OUT
BY ANALYZING THE
INK THAT WAS USED
TO PRINT IT?

PHOTO: Courtesy of Carolina
Biological Supply Company

INTRODUCTION

Most of the mixtures you have examined in this unit have consisted of
two substances. For example, the solutions you studied were mixtures
of a solvent and a solute. Different solvents can be used to remove more
complex mixtures, or stains. The success of stain removal depends on
the solubility of the pure substances that make up the stains. Ink is an
example of a stain that is frequently given to the dry cleaners to remove.
The dry cleaner must find a solvent that dissolves the ink solute to
remove it. In this lesson, you will take a closer look at what happens
when solvents are added to inks, and how some of these properties can
be used to identify inks from different sources.

OBJECTIVES FOR THIS LESSON

Use paper chromatography to analyze and identify inks.

Apply paper chromatography techniques to "solve a crime."

MATERIALS FOR LESSON 4

For you

1	Copy of Student Sheet 4.1: Analyzing Inks
1	copy of Student Sheet 4.2: Comparing Inks
1	copy of Student Sheet 4.3: Identifying Inks

For you and your lab partner

1	250-mL beaker
2	pieces of chromatography paper
1	pencil
1	metric ruler
	Access to water

For your group

1	brown marker
1	green marker
1	red marker
3	black markers (labeled B, C, and D)
	Access to a wall clock

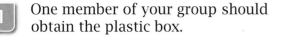

GETTING STARTED

1 One member of your group should obtain the plastic box.

2 Check the contents of the plastic box against the materials list and remove the apparatus for your pair. (In this lesson, you will work with a partner, but you will share the markers with the other pair in your group.)

3 Put about 25 mL of water into the 250-mL beaker.

4 Quickly (for about 1–2 seconds) dip the tip of the green marker into the water.

5 Observe what happens. Answer this question on Student Sheet 4.1: Analyzing Inks: What happened when you put the tip of the green marker into the water?

6 Based on your observations, what do you know about the ink? Record your answer.

7 Your teacher will lead a short discussion about the observations you have made.

INQUIRY

ANALYZING INKS

PROCEDURE

1 Empty the beaker you used in "Getting Started."

2 Put about 50 mL of water into the beaker.

3 Place a spot of green ink (about 2 mm in diameter) on one piece of chromatography paper, as shown in Figure 4.1.

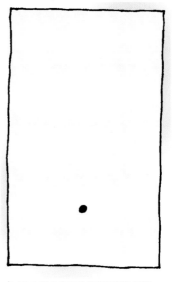

▶ **PLACE A SPOT OF INK IN THE POSITION SHOWN.**
FIGURE **4.1**

Fold over the paper about 3 cm from the top.

 Rest a pencil across the top of the beaker containing the water. Hang the paper on the pencil, as shown in Figure 4.2, for 5 minutes. Make sure the spot of ink is above the surface of the water.

6 Observe what happens to both the water and the ink. After 5 minutes, remove the paper from the beaker.

7 Draw your results on the "chromatography paper" on Student Sheet 4.1: Analyzing Inks. Under Step 4, describe what you observed.

8 What can you conclude about the ink in the green marker? Record your ideas.

9 Contribute your ideas to a short class discussion.

10 Read "Introducing Paper Chromatography" on the next page.

▶ USE A PENCIL TO HANG THE PAPER IN THE BEAKER
CONTAINING THE WATER. MAKE SURE THE INK
SPOT IS ABOVE THE SURFACE OF THE WATER.
FIGURE 4.2

COMPARING INKS

PROCEDURE

1 How could you use chromatography to compare the composition of the inks in the green, red, black (Pen B), and brown markers? You will need to make a fair comparison. Discuss this question with your partner. Your teacher will conduct a brainstorming session to come up with a class approach to this problem.

2 Use the procedure agreed on in class to compare the four inks. While you are waiting for the inks to move up the paper, write and/ or illustrate the procedure under Step 1 on Student Sheet 4.2: Comparing Inks.

3 After the inks have moved up the paper, look at the chromatograms. Answer the following questions on your student sheet:

- What can you say about the composition of each color?

- Do dyes of the same color always behave in the same way?

- What does this tell you about these particular dyes?

READING SELECTION

BUILDING YOUR UNDERSTANDING

INTRODUCING PAPER CHROMATOGRAPHY

The technique you used to separate the dyes in the green ink is called paper chromatography. The patterns produced on the paper are called a chromatogram. Chromatography works because the ink you used consists of several dyes of different colors that are dissolved in water. Each dye is a different solute and moves at a different speed through the chromatography paper. Just like cars traveling at various speeds on a highway, the faster solutes move ahead of the others and eventually become separated from them. The speed at which each solute moves through the paper depends on the solubility of the solute in the solvent (water). The more soluble dyes move faster than the less soluble ones. ■

IDENTIFYING INKS

PROCEDURE

1 Chromatography is a separation technique that is used for many purposes, including solving crimes. Read "The Case of the Unidentified Ink" on the next page. Then apply your knowledge of chromatography to find out whether a crime has been committed and, if so, how to solve it.

2 Discuss with your partner how you are going to determine whether a crime has been committed and, if so, who committed it.

3 Outline the procedure you will follow on Student Sheet 4.3: Identifying Inks.

4 Make sure your teacher approves your plan, and then conduct your investigation using the sheets of chromatography paper prepared by your teacher.

5 Record the results of your investigation in Step 2 on Student Sheet 4.3 as a description, diagram, or chromatogram.

6 Write a summary of your conclusions that can be presented in court.

7 Empty your beakers and return all of the materials to the plastic box.

REFLECTING
ON WHAT YOU'VE DONE

1 Discuss your chromatography findings with the class. Some pairs will be asked to present their summaries.

2 Members of the jury in this case have no knowledge of chromatography. It is your job to educate them about the process. In your science notebook, write a paragraph explaining what chromatography is, and another describing how the process works.

READING SELECTION

BUILDING YOUR UNDERSTANDING

THE CASE OF THE UNIDENTIFIED INK

JUAN TUTRIFORE
6543 21st Street
Hiccup, Texas 78987

$\frac{16-4}{257}$/2486   6789

Date *Mar 23, 2000*

Pay to the Order of *Ludmilla Hotso* $ *723,000.* 00

Seven Hundred Twenty-Three Thousand Dollars

Unnatural
BANK OF HICCUP, N.A.

For *Down payment*

⑈5880⑈700⑈: 899 741 9563⑈⑈ 6243

▶ EXHIBIT A: IS THIS CHECK A FORGERY?

Forensic scientists help solve crimes by using a wide range of scientific techniques to gather evidence. Chromatography is one important technique that they use to catch criminals. In this inquiry, you will need to place yourself in the role of just such a scientist.

Exhibit A shows a check with a signature on it. Is the signature on the check a forgery? Was the check written by the owner or by someone else?

There are three suspects in this case, each of whom uses a different pen. The only way to find out who signed the check is to analyze the ink in the signature and compare it with the ink in the pens handed over as evidence (Exhibits B, C, and D).

Pen B belongs to the owner of the checkbook, pen C belongs to a notorious criminal, and pen D belongs to a teller at the bank. Your lab assistant has already extracted some ink from the signature on the check and placed a spot of it on a piece of chromatography paper.

Who wrote the check? Was a crime committed? ■

"Separation Science" at the FBI

▶ WHAT DOES THE HEADQUARTERS OF THE FBI IN WASHINGTON, D.C., HAVE TO DO WITH CHROMATOGRAPHY?

PHOTO: National Science Resources Center

Two thieves rob a small-town bank of thousands of dollars. A bomb explodes in Oklahoma City, killing more than 150 people.

These two crimes are very different, but one of the techniques that crime detection experts use to investigate them is the same. It's called chromatography. One of the largest chromatography laboratories for crime detection is located in Washington, D.C., at the headquarters of the Federal Bureau of Investigation (FBI). Samples from crime scenes across the nation are sent to the FBI for forensic analysis. (Forensic analysis is a special type of scientific analysis that is performed in conjunction with legal or court proceedings.)

Many experts in forensic analysis work at the FBI. Some spend their time comparing fibers and hairs from crime scenes; others compare the soil from a suspect's shoes with the soil found at a crime scene. There are many rooms full of special equipment. Two of those rooms are dedicated specifically to chromatography. It is here that FBI Special Agent Kelly Mount works.

"Chromatography," explains Kelly Mount, "is a 'separation science.' It's the technique we use in our labs to separate the components of a mixture." Chromatography has many uses; it's not limited to crime detection. In fact, the first chromatography was done by a Russian botanist in 1906. He discovered that the pigments extracted from green leaves into a solution of ether, including chlorophyll pigment, could be separated from one another by passing the solution through a tube containing powdered calcium carbonate (chalk).

Mount uses a variety of substances to separate mixtures. Gases and liquids are the most common. Separating the substances, Mount says, is just the beginning of forensic analysis. "The purpose of chromatography is not to identify, but to separate," she emphasizes. (The identification comes after the separation.) Once scientists have their results, they have to compare them with the results of other analytical techniques before they can be admitted into evidence.

READING SELECTION
EXTENDING YOUR KNOWLEDGE

▶ CHROMATOGRAPHY HAS MANY APPLICATIONS BEYOND SOLVING CRIMES. HERE, A NUTRITIONAL BIOCHEMIST USES HIGH-PERFORMANCE LIQUID CHROMATOGRAPHY TO MEASURE CAROTENOIDS IN BLOOD SAMPLES.

PHOTO: Stephen Ausmus, Agricultural Research Service/U.S. Department of Agriculture

"Let's go down the hall and take a look at two ways we use chromatography," Mount says. "The first one is simple and inexpensive—students do something similar to this in science class. The other one needs some sophisticated equipment.

"First, we'll take a look at how we use a simple form of chromatography, called thin-layer chromatography, to help track down bank robbers," she says. "Here's how it works. When banks bundle paper currency together, they routinely include a special security device inside some of the packs. This device has a miniature 'bomb' inside of it. When triggered, the bomb explodes. It doesn't do any damage to humans, but it does release a bright red liquid. The liquid is impossible to wash out."

When a robber says, "Hand over the cash," the bank teller obligingly turns it over, making sure to include a bundle of the specially packaged money. Soon after the robber leaves the bank, the device explodes, showering his or her car, clothing, or bag with the distinctive red dye.

The suspected criminal is sometimes caught "red handed." A sample of clothing or other material stained with the dye is sent to Kelly Mount's unit for analysis. Is it the dye from the security device? Or did the red color come from another source?

The chemical composition of the red dye used by banks is unique. No other dye has the same composition. Mount compares this dye of known composition with the dye found on a suspected criminal. She takes a crime scene sample of the material containing the dye and puts it in a solvent to extract the dye from the material. Mount then places several drops of the dissolved sample and the standard dye along a line at the bottom of a small, specially coated plate. (The thin coating on the plate is what gives thin-layer chromatography its name.) She stands the plate upright, with the samples along the bottom, in a container with a small amount of liquid. As the drops become moist and interact with the coating on the plate, they begin to move up the plate at different rates, depending on the solubility of the components. Once the liquid gets near the top of the plate it is removed from the container. Then, the positions of the drops are compared. All dyes made from the same components will form the same pattern on the plate. So, if the pattern of the crime scene sample matches the pattern of the dye used by banks, another crime has been solved.

BOMBS AND EXPLOSIVES

Chromatography also comes in handy for analyzing the materials used in bombs and explosive devices. The FBI analyzes samples from all major bombings involving the United States, including the one at the Murrah Federal Building in Oklahoma City and others causing airline crashes. The technique is called high-pressure liquid chromatography—HPLC, for short.

The first step is examination under a microscope. "Most bomb samples look pretty much alike. They look like black powder," says Mount. Even so, this first step is important. The scientists might, for example, be able to sort out small pieces of material from the residue.

The next step is extraction. The chemists place the sample in a solvent such as water. Once in solution, the particles in the sample may, depending on the composition of the sample, separate into smaller particles that carry positive or negative charges.

A small amount of the solution is placed in the HPLC machine. It moves up to the top, where it mixes with another liquid, and is then forced downward under pressure through a narrow glass column that is filled with a porous substance.

What happens in the column is the critical step. "Some of the [particles]," explains Mount, "seem to like it better in the tube than others. They stay longer."

The speed at which the particles leave the column is recorded by a detector, which then prints out the information. By comparing the time that the particles have stayed in the column with known retention times, Mount and her colleagues are able to distinguish the various types of particles in the test sample.

STILL A LOT TO LEARN

Does it always work? "No," says Mount. "Sometimes we find nothing. And other times, we find nothing conclusive. It's also important to note that when it comes to explosive materials, HPLC is only a qualitative analysis technique. It helps us identify what materials are in an unknown powder. It doesn't provide quantitative information; in other words, we can't tell how much of each substance is in the powder."

Kelly Mount loves her work. To prepare for her career, she earned a bachelor's degree in chemistry and then went on to get a master's degree in forensic science. Life in the lab is never routine—this means getting called into the lab on weekends or even at night when there is an emergency. Whether the problem is bombs or banks, Mount has the expertise to help the FBI solve its mysteries. ■

DISCUSSION QUESTIONS

1. Envision a crime scene not described in the reading selection in which chromatography could be useful. Describe how chromatography could be used to help solve the crime.

2. Describe the procedure you would use to find out whether carrots get their color from one or many substances.

CHANGING MIXTURES

INTRODUCTION

Most of the materials used to make things are mixtures. In fact, if you look around your classroom, you will discover that it is very difficult to find any pure substances. Concrete, bricks, paper, wood, steel, and glass are all mixtures. Materials can be made to have specific properties by altering the types and the amounts of substances that go into a mixture. For example, changing the amount of water or the amount and type of aggregate (stones) in concrete will change its strength when it is set. The wetter the original mix, the weaker the concrete. Adding different kinds of aggregate can also change concrete strength.

A solution is a special kind of mixture. Can a solution be used to make things? Can the properties of a solution be altered? What happens to the properties of a solvent when you add a solute to it? Are the properties of a solution different from those of the solvent and the solute (or solutes) from which it is made? Does the amount of solute you add to a solution affect the properties of the solution? You may be surprised by the answers you find to these questions during the course of this lesson.

▶ THIS SKYSCRAPER IS BEING BUILT USING A WIDE VARIETY OF MIXTURES, INCLUDING CONCRETE, STEEL, AND GLASS.

PHOTO: Joi Ito/creativecommons.org

OBJECTIVES FOR THIS LESSON

- Measure the effect of different quantities of salt on melting and boiling points.

- Compare the melting points of different alloys.

- Discuss the technological applications of solutions and other mixtures.

▶ MATERIALS FOR LESSON 5

For you

1	copy of Student Sheet 5.1: Adding Salt to Ice
1	copy of Student Sheet 5.2: Adding Salt to Boiling Water
1	copy of Student Sheet 5.3: Investigating Solid Solutions
1	pair of safety goggles

For your group

1	plastic spoon
1	black marker
2	250-mL beakers
2	thermometers
1	aluminum pan
1	jar of sodium chloride (common salt)
1	piece of solder, color-coded blue
1	piece of solder, color-coded green
1	piece of solder, color-coded red
1	burner
1	burner stand with gauze
	Crushed ice
	Access to hot water
	Access to a clock or watch with a second hand

GETTING STARTED

 1 During this lesson, you will work in a group of four. Discuss the following questions with the members of your group:

A. Can you think of at least one mixture that has the properties of both of the substances from which it is composed?

B. Can you think of at least one mixture that has the properties of only one of the substances from which it is composed?

C. Can you think of at least one mixture that has properties completely different from the properties of the substances from which it is composed?

Record your answers in your science notebook.

2 Be prepared to contribute to a class brainstorming session about these questions.

 **SAFETY TIP**

Wear your safety goggles throughout the lesson.

INQUIRY

ADDING SALT TO ICE

PROCEDURE

1 One member of your group should collect the plastic box containing the materials. Label one of the beakers "A" and the other "B."

2 Collect the crushed ice from your teacher.

3 Fill Beaker A and Beaker B one-third to one-half full with crushed ice.

4 Place a thermometer in each beaker. Measure the temperature of the crushed ice for each. Record your results on Student Sheet 5.1: Adding Salt to Ice.

5 Discuss with other members of your group what you think will happen to the ice in Beaker A if you add two heaping spoonfuls of salt to it. What will happen in Beaker B without any salt? Record your predictions.

6 Add two spoonfuls of salt to Beaker A only. Use the plastic spoon to stir the mixture. Observe each beaker and measure the temperature of each. What happened to the ice in Beaker A when you added the salt? What happened in Beaker B? Record your observations.

7 Write what you think will happen if you add another two spoonfuls of salt to Beaker A. What will happen in Beaker B without the salt?

8 Add the salt as before and record your observations.

9 Record your answers to the following questions:

- What effect does salt have on the state of matter of ice?

- What effect does salt have on the melting point of ice?

- What effect does adding more salt have on the temperature of the ice/salt mixture?

10 Does the amount of salt you add affect the temperature or melting point of the ice? Record your answer. Participate in a class discussion of your results.

11 After the discussion, dispose of the contents of both beakers by pouring them down the sink. Rinse the beakers and the thermometers and return the materials to the plastic box.

CEMENT IS A MIXTURE OF WATER, POWDERY CLAY, LIMESTONE, AND SAND OR SHALE. HOW IS IT DIFFERENT FROM THE SUBSTANCES FROM WHICH IT IS MADE?

PHOTO: Midtown Crossing at Turner Park/ creativecommons.org

INQUIRY 5.2

ADDING SALT TO BOILING WATER

PROCEDURE

1 Put about 100 mL of hot water in a 250-mL beaker. Place a thermometer in the beaker.

 **SAFETY TIPS**

Carefully follow your teacher's instructions for the use of burners.

Tie back long hair.

Be careful when handling hot objects.

2 Follow the procedure outlined by your teacher for igniting the burner.

3 Place the thermometer and beaker on the stand and heat the water until it boils. Answer the following questions on Student Sheet 5.2: Adding Salt to Boiling Water:

- How can you tell the water is boiling?

- At what temperature does the water boil?

4 Predict what will happen to the boiling point of the water if you add two heaping spoonfuls of salt to it. Record your prediction.

5 Add two spoonfuls of salt to the water. Watch the boiling water and the thermometer. Record your observations.

6 Add two additional spoonfuls of salt to the boiling water. Record your observations.

7 Record your answers to the following questions:

- What effect did adding salt have on the boiling point of water?

- Does the amount of salt added affect the temperature or the boiling point of water?

8 Write a short paragraph summarizing the effect salt has on the melting and boiling points of water.

9 Read "Changing Melting and Boiling Points," then participate in a class discussion.

CHANGING MELTING AND BOILING POINTS

PHOTO: © David Marsland

Do you know what is inside the strange building shown in this photograph? The building is used to store salt that is spread on the road when weather conditions are below the freezing point of water. What effect will the salt have on icy roads? Why store the salt in a building?

PHOTO: William Billard/creativecommons.org

This truck removes snow and spreads salt. How does spreading salt on roads reduce accidents?

PHOTO: © David Marsland

A solution of antifreeze is used to fill car radiators. It lowers the freezing point of the water in the radiator of the car. It also raises its boiling point and reduces corrosion. Why are the properties of this solution useful to motorists? Why isn't salt used instead? ■

INQUIRY **5.3**

INVESTIGATING SOLID SOLUTIONS

PROCEDURE

1 In this inquiry, you will investigate how impurities affect the melting point of three metal mixtures called solders. Because these solders melt at temperatures above the range of the thermometers you will use, you will measure the amount of time it takes for each of the solders to melt. You will then compare these measurements to determine the melting points of the solders. Your teacher will demonstrate how the apparatus in this inquiry should be used. Watch carefully and then read the instructions and Safety Tips before you start.

2 Assemble the apparatus as shown in Figure 5.1, but do not place the burner into position under the stand until Step 6.

SAFETY TIP

Solder is toxic if ingested. Do not put solder in your mouth.

3 Place the pieces of solder on the aluminum pan in the same positions as those shown in Figure 5.2.

4 Make sure the aluminum pan is positioned in the center of the gauze. Make sure you know where each color-coded piece of solder is by completing the diagram in Step 1 on Student Sheet 5.3: Investigating Solid Solutions. Warning: As the solder gets hot, the color codes may disappear.

5 Follow the procedure outlined by your teacher for igniting the burner. (If you are using a Bunsen burner, use the gas tap to make a flame about 4 cm high, and open the air hole about halfway.)

▶ HOW TO ASSEMBLE THE ALCOHOL BURNER APPARATUS FOR INQUIRY 5.3. (YOUR BURNER MAY DIFFER FROM THIS ONE.) DO NOT PLACE THE BURNER INTO POSITION UNDER THE STAND UNTIL STEP 6.

FIGURE **5.1**

PHOTO: © 2009 Carolina Biological Supply Company

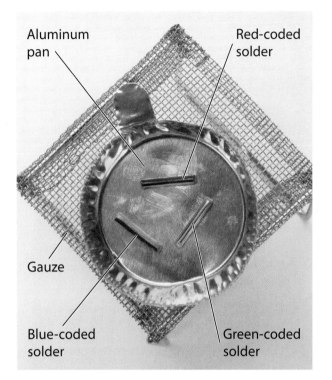

Aluminum pan

Red-coded solder

Gauze

Blue-coded solder

Green-coded solder

▶ PLACE THE PIECES OF SOLDER ON THE ALUMINUM PAN.
FIGURE **5.2**
PHOTOS: © 2009 Carolina Biological Supply Company

SAFETY TIPS

Do not lean over the apparatus during heating.

Observe the solders from a distance of 0.6–1 meter (2–3 feet).

Inquiry 5.3 continued

6 Using a clock or watch with a second hand, start recording the time at time 0. Move the burner so that the top of the flame is exactly in the center of the aluminum pan. Stand 0.6-1 m (2-3 feet) away from the apparatus.

7 Watch the pieces of solder very carefully because changes may happen very quickly. Record in Table 1 on Student Sheet 5.3 when each piece of solder begins to show signs of melting (change in shape, seems more liquid than solid, etc.). Stop heating when all the pieces have melted, or after 5 minutes (whichever comes first). If any of the pieces remains unmelted, record this information in Table 1.

8 Extinguish your burner.

SAFETY TIPS

Do not touch your apparatus for at least 5 minutes. It is still very hot.

Wash your hands before leaving the lab.

9 Discuss the following questions with your group members. Once your group has agreed on the answers, write them under Steps 3 through 8 on Student Sheet 5.3.

- How do you think the length of time to melt relates to the melting points of these solders? (Obtain the melting points from your teacher and add them to Table 1.)

- Did all of the solders melt at the same temperature?

- What effect do other metals have on the melting point of tin? (Use the information in Table 1 on Student Sheet 5.3 to help you answer this question.)

- What effect does a greater amount of silver have on the melting points of the mixtures?

- Look at the melting points of the pure metals in Table 1. Three of these metals are very difficult or impossible to melt with a lab burner. Based on this information, do you think that mixtures of metals always show a combination of the properties of their components?

- Why are the low melting points of these alloys a useful property for solder?

10 Make sure your apparatus is cool. Return the materials to the plastic box and give any unused pieces of solder to your teacher.

BUILDING YOUR UNDERSTANDING

ABOUT ALLOYS

Alloys are mixtures that contain at least one metal. Most alloys are solid solutions. Although many alloys consist of two metals mixed together (for example, silver and tin in some solders), the most widely used alloy is the mixture of carbon and iron that is called steel. Steel is much stronger than pure iron. Its properties can be changed by adding other substances to it. For example, manganese makes steel harder, and chromium, which is used to make stainless steel, stops steel from rusting.

Alloys of other metals, such as aluminum and titanium, provide the high-strength, low-density materials needed to make aircraft. Alloys of tungsten and cobalt are used in materials that must resist the effects of high temperature (for example, rocket engines). Bronze is an alloy of copper and tin. Pure gold is very soft, causing items made from it, such as jewelry, to be easily damaged. Therefore, gold is often alloyed with silver and copper to produce a harder metal. You will learn more about some of these metals in Lessons 7 and 8. ■

REFLECTING
ON WHAT
YOU'VE DONE

1. Your teacher will lead a class discussion on your results for Inquiry 5.3. Use the data and answers on Student Sheet 5.3 to participate in this discussion.

2. Read "About Alloys."

3. Participate in a class discussion on alloys, how their properties are manipulated, and how they are used.

THE SAMURAI'S SWORD

The properties of metal objects are determined not only by all the different metals that make them up but also by the way the metals are mixed together and treated. For thousands of years, metal workers, or smiths, have been altering the properties of metals by heating, hammering, and using other treatments to make objects as diverse as springs and gun barrels. The famous swords of the Samurai warriors of medieval Japan are one example of how smiths used the properties of particular metals for specific purposes.

The first Samurai were soldiers who were hired by landowners to protect their property from bands of robbers. From the 12th to the 19th century, even though Japan had emperors, the Samurai actually ruled Japan. The sons of Samurai were trained from early childhood for careers as warriors. A young man began his career at about age 15, when he received his first sword in a special ceremony.

Although each Samurai also carried a bow and arrow (and was trained in wrestling and judo), sword fighting was his most important skill. And the Samurai's swords were special indeed. Each Samurai had a long sword and a short one. The long sword, called the *katana*, was his main weapon. Its steel blade was designed to kill an enemy with one swipe!

To make a *katana*, a swordsmith used two types of steel. The core of the sword was made of soft, flexible, low-carbon steel (an alloy of iron with a little carbon). The jacket, or outer part of the sword, was made of hard steel that contained a greater proportion of carbon than the core. The combination of these two kinds of steel gave the sword the flexibility to withstand a hard blow and a hard, razor-sharp edge that would not be dulled during battle.

横川勘平藤原宗則

徒頭　禄　廿五両五人扶持

芝新銭座森鷹の家中ふ伯父ありふれが
らの家に同居るー日に堀部安兵衛が
本所の宅にいて万変ととりさうるひ
討への廿ろ要用の道具を取あろめさし
支ろ化やう心を用ひり本意を達して後
死ふのぞころくすくちがー死出の
延遙ろふあろがなも被先得て乃志ふちん

法号

刅常水劍信士
行年
三十七才

READING SELECTION
EXTENDING YOUR KNOWLEDGE

► KENJI MISHINA, A GREAT
MASTER JAPANESE SWORD
POLISHER, CAREFULLY
POLISHES A SAMURAI SWORD.

PHOTO: © Kenji Mishina

The swordsmith treated both steels with techniques that improved the performance of the sword even further. He began by heating a lump of raw low-carbon steel—about the size of a brick—in a forge (a furnace that burns charcoal at very high temperatures). The swordsmith then hammered the steel on an anvil until it was flat. Then he folded it in half crosswise and hammered it out again. He repeated this process many times to drive out any impurities from the metal. Finally, he shaped it into a long, thin wedge.

Next, the swordsmith began to work on the high-carbon steel. He followed the same process as the one he used for the low-carbon steel, but this time, he hammered and folded many more times. The final piece of metal had up to 30,000 layers. The swordsmith made the jacket somewhat longer than the core.

Next, he joined together the two parts of the blade. The jacket was wrapped around the core, and the swordsmith heated and hammered the two pieces until they formed a solid bond. He had to be extremely careful; if an air bubble or piece of dirt remained between the two parts of the blade, the sword would be worthless in battle.

The blade was then tempered, a process that is used to control the properties of the steel. The blade was heated and then cooled by being plunged into water. The swordsmith coated the sword with clay to control the cooling process. Where the coat of clay was thick, the steel would cool more slowly, and this would make it flexible. The edge of the blade was given a thin coat of clay, which allowed it to cool very quickly, a process that made the edge even harder.

The swordsmith sharpened and polished the blade. The layers, or grain, were visible on the shiny surface. Finally, he tested the blade—on iron sheets, armor, and sometimes, the bodies of executed criminals.

The Samurai swords were deadly but beautiful. The blades were decorated, and the handles were inlaid with pearls and other jewels.

Samurai swords were passed from generation to generation. Upon reaching manhood, a son received his father's sword, along with stories of the brave acts that had been accomplished with it. ■

DISCUSSION QUESTIONS

1. Fold a large piece of paper in half. Turn the paper, and fold it in half again with the new fold perpendicular to the old fold. How many layers do you have? How many times can you fold the paper? Could you make more folds with a larger piece of paper? What relationship can you see between the number of folds and the number of layers? Challenge a classmate to make at least eight folds in a piece of paper.

2. What techniques did swordsmiths use to modify the properties of different parts of a *katana* blade? Investigate how these techniques are applied for other purposes today.

Ice Cream in the Old Days

Which came first, ice cream or freezers? Everyone knows that a freezer is needed to store ice cream. To keep ice cream solid, it has to be stored well below the freezing point of water (0°C or 32°F). Making ice cream also requires the same low temperatures. People didn't have freezers in the old days, so were they able to make ice cream?

The answer is yes. Most people used to make their own ice cream at home. They would have ice delivered to their house by an ice-making company, or they would use ice they had collected in winter and stored underground. They would start by making an ice cream mixture. They would combine the mixture in a metal container (one that's good at conducting heat) and then place the container in a bucket containing crushed ice and a little water.

Next, they added salt to the ice. The ice would immediately start to melt. To melt, ice takes in heat from its surroundings, cooling down the container of the ice cream mixture to below the freezing point of water. They would continuously stir the ice cream mixture so it produced small ice crystals, which gives ice cream its creamy texture. Sometimes they would have to use as much as a pound of salt to make the ice cream.

You can make your own ice cream by following Great Grandma's vanilla ice cream recipe. After you make it, you can add your favorite toppings. ■

▶ DELICIOUS! BUT HOW WAS ICE CREAM MADE WITHOUT A FREEZER?

PHOTO: Library of Congress, Prints & Photographs Division, FSA/OWI Collection, LC-USF33-016109-M3

Vanilla Ice Cream

1 cup of sugar
2 cups of milk
1 cup of table cream
1 cup of whipping cream
2 teaspoons of vanilla

1. Pour the milk and table cream into a saucepan.
2. Bring the milk and cream to a boil.
3. Beat sugar into the heated milk and cream.
4. Allow the mixture to cool. Pour the mixture into a bowl.
5. Stir in the whipping cream and vanilla.
6. Place the bowl in a bucket containing crushed ice. Stir lots of salt into the ice (make sure no salt gets into the ice cream mixture).
7. Continuously stir the ice cream. Keep adding salt to the ice until all the ice has melted.
8. EAT!

▶ THIS ICE CREAM MACHINE REQUIRED THE USE OF ICE AND SALT TO LOWER THE TEMPERATURE OF THE ICE CREAM MIXTURE TO THE POINT WHERE IT WOULD FREEZE.

PHOTO: Library of Congress, Prints & Photographs Division, FSA/OWI Collection, LC-USF33-031141-M4

? DISCUSSION QUESTIONS

1. What where the pros and cons of the way people got ice before households had refrigerators?

2. Some brands of ice cream are "slow churned," and have smaller ice crystals than the usual ice cream. Find out how the "slow-churned" process is related to the process described in the reading selection.

BREAKING DOWN A COMPOUND

INTRODUCTION

In your previous work, you looked at the characteristic properties of pure substances. Then you investigated how those properties can differ from the properties of mixtures. Now you will focus on two groups of pure substances known as elements and compounds. In this lesson, you will examine the composition of the pure substance you have encountered most often during the course of this unit—water. You know that water has several characteristic properties that can identify it as a single substance rather than a mixture. These properties include its appearance, density, melting and boiling points, and ability to dissolve a wide range of solutes.

You will investigate what happens to water when electricity is passed through it. Sometimes, passing electricity through a liquid can give you clues about the composition of the liquid. If water is a pure substance, why try to find out its composition? Do the inquiry and find out what happens.

THE SURFACE OF THE EARTH IS FOUR-FIFTHS WATER. WHAT IS WATER MADE FROM?

PHOTO: NASA

OBJECTIVES FOR THIS LESSON

Conduct an experiment to determine what happens when electricity is passed through water.

Investigate some physical and chemical properties of gases.

Discuss the differences between compounds and elements.

▶ MATERIALS FOR LESSON 6

For you

1	copy of Student Sheet 6.1: Electrolysis of Water
1	pair of safety goggles

For your group

1	plastic container
2	test tubes
1	electrode stand
1	jar of sodium sulfate
1	lab scoop
1	plastic spoon
1	wooden splint

For your group and another group to share

2	6-V batteries
1	insulated connector wire with alligator clips
	Access to water
	Access to a burner

GETTING STARTED

1 You often hear the terms "element" and "compound." In fact, they are used in the title of this unit. What do you think these terms mean? Without referring to a dictionary or the glossary, write definitions for each of these terms in your science notebook.

SAFETY TIPS

Wear safety goggles throughout the inquiry.

Tie back long hair.

2 Next, list two examples of an element and two examples of a compound.

3 Participate in a brainstorming session on the meanings of these terms. You will look at these ideas again at the end of the lesson.

▶ SAND IS A MIXTURE. WATER IS A COMPOUND. WHAT IS THE DIFFERENCE?

PHOTO: Randolph Femmer/National Biological Information Infrastructure

SPLITTING WATER

PROCEDURE

1 Your teacher will demonstrate the procedure for passing an electric current through water. After the demonstration, follow Steps 2 through 10 to set up your apparatus.

2 Place the electrode stand in the plastic container.

3 Make sure the leads hang over the side of the container.

4 Add water to the container so that the tips of the electrodes are covered by about 1 cm of water (see Figure 6.1).

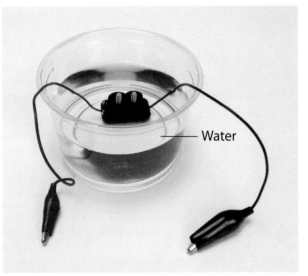

▶ **PLACE THE ELECTRODE STAND IN THE PLASTIC CONTAINER AND ADD WATER.**
FIGURE **6.1**
PHOTO: © 2009 Carolina Biological Supply Company

5 Add two lab scoops of sodium sulfate to the water.

6 Stir the solution with the plastic spoon.

7 Submerge one of the test tubes in the container of sodium sulfate solution (see Figure 6.2).

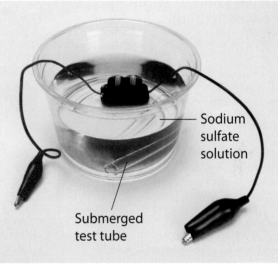

▶ **SUBMERGE THE TUBE AND MAKE SURE IT IS FULL OF SODIUM SULFATE SOLUTION.**
FIGURE **6.2**
PHOTO: © 2009 Carolina Biological Supply Company

8 When the tube is full of solution, place your thumb or finger over its top. Making sure the open end of the tube is below the liquid, place the opening of the tube over one of the electrodes. The tube must still be full of liquid.

Inquiry 6.1 continued

9 Repeat Steps 7 and 8 with the second test tube. (Do not worry if a few small bubbles of air get into the tubes.) Figure 6.3 shows what your apparatus should look like. Wash your hands after handling the tubes.

10 Clip the connector wires onto the batteries. Clip the red connector wire to the positive terminal of one battery and the black connector wire to the negative terminal of the other battery. Make sure the two remaining terminals on the batteries are connected (see Figure 6.4). You will be sharing the two batteries with another group. Figure 6.5 shows how to set up the apparatus for both groups.

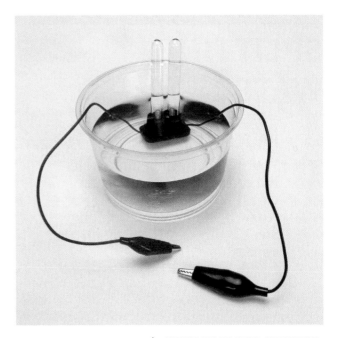

▶ AT THIS STAGE, YOUR APPARATUS SHOULD LOOK LIKE THIS.
FIGURE **6.3**
PHOTO: © 2009 Carolina Biological Supply Company

▶ CLIP THE CONNECTOR WIRES ONTO THE BATTERIES AS SHOWN.
FIGURE **6.4**
PHOTO: © 2009 Carolina Biological Supply Company

▶ YOU ARE SHARING BATTERIES WITH
ANOTHER GROUP. JOINTLY, YOUR
APPARATUS SHOULD LOOK LIKE THIS.
FIGURE **6.5**
PHOTO: © 2009 Carolina Biological Supply Company

 11 If you have assembled your apparatus correctly, you should soon start to see something happening near the electrodes. What do you observe happening at each of the electrodes? Record your observations on Student Sheet 6.1: Electrolysis of Water.

12 What do you observe about the volumes of the two substances collected in the tubes? Record your answer.

13 Participate in a class discussion about your observations.

Inquiry 6.1 continued

14 You (or your teacher, if you do not have time to collect enough of each gas) can test the gases produced in each test tube by putting a burning splint into each tube. If your group is conducting this test, follow these instructions:

SAFETY TIPS

Carefully follow your teacher's instructions for handling the burning splint.

Be careful when handling hot objects.

A. Disconnect the battery from your apparatus.

B. Have one member of your group ignite the splint so it is burning.

C. Have another group member carefully remove a gas-filled tube from the negative electrode. It is all right if a little solution in the tube empties back into the plastic container. Hold the tube with the open end down (inverted).

D. With the open end down, tilt the test tube at a 45° angle and quickly put the lit end of the splint into the mouth of the test tube. Look and listen.

E. Record your results under "Tube 1" in Table 1 on Student Sheet 6.1.

F. Refill the test tube with sodium sulfate solution, reconnect the battery, and continue to collect gas.

G. When the test tube over the positive electrode is filled with gas, perform the same test.

H. Record the results under "Tube 2" in Table 1.

15 When the tube over the negative electrode has refilled with gas, perform the following test:

A. Have one member of your group remove the test tube from the negative electrode, keeping it sealed with a thumb or finger. Hold the tube in an upright position (open end up).

B. Another member of the group should ignite the splint.

C. Blow out the splint and quickly put the glowing end into the tube of gas.

D. Carefully observe what happens and record the results under "Tube 3" in Table 1.

E. Repeat the test with a tube full of gas from the positive electrode.

F. Record your results under "Tube 4" in Table 1.

16 Hydrogen gas burns with a squeaky pop, and oxygen relights a glowing splint (or makes it glow much brighter). Based on your experimental evidence, what is inside each tube? Write your answers under Steps 4a and 4b on Student Sheet 6.1.

17 Read "The Electrolysis of Water."

READING SELECTION

BUILDING YOUR UNDERSTANDING

THE ELECTROLYSIS OF WATER

In Inquiry 6.1, electricity was used to split water into hydrogen and oxygen. This process is called electrolysis ("electro" refers to electricity, and "lysis" means "to break apart"). To break down water through the process of electrolysis, electricity must be able to flow through the water and complete an electrical circuit.

Electricity does not flow easily through pure water because pure water is a poor conductor of electricity, making it difficult to complete an electrical circuit. Adding sodium sulfate to pure water helps the water conduct electricity, which makes it easier to complete an electrical circuit. The energy from the electric current causes a chemical reaction to take place. During the reaction, sodium sulfate does not produce any products that can be detected. All of the gases produced during the electrolysis of water come from the water. ∎

REFLECTING
ON WHAT
YOU'VE DONE

❶ Write the answers to the following questions on Student Sheet 6.1:

A. Which two gases make up water?

B. You know that water is a pure substance. You have found out that it is made from two gases. Both gases are also pure substances. However, these gases cannot be broken down into other substances. Pure substances that cannot be broken down into other substances are called elements. Pure substances that are made up of more than one element are called compounds. Do you think water is an element or a compound?

C. Unlike mixtures, pure substances that are compounds always have the same ratio of elements in them; in other words, they have fixed compositions, or formulas. What is the ratio of hydrogen to oxygen in water?

D. Water is sometimes written as a formula, H_2O. What do you think this formula means?

❷ Think about how the characteristic properties of water differ from the characteristic properties of hydrogen and oxygen. Read about some of them in the reading selection "Hydrogen and Oxygen" on pages 66-67. Your teacher will collect your ideas and compile them in a table. Copy this table into your science notebook.

❸ Review the definitions of the terms "element" and "compound" that you wrote at the beginning of the lesson. Discuss with your partner how your ideas have changed. Write your ideas in your science notebook.

READING SELECTION

BUILDING YOUR UNDERSTANDING

HYDROGEN AND OXYGEN

▶ WHEN FLAMMABLE GASES SUCH AS ACETYLENE ARE BURNED IN PURE OXYGEN, VERY HIGH TEMPERATURES RESULT. THIS OXYACETYLENE TORCH BURNS A MIXTURE OF ACETYLENE GAS AND OXYGEN, WHICH PRODUCES A FLAME HOT ENOUGH TO CUT OR WELD STEEL.

PHOTO: DoD photo by Cpl. Bryson K. Jones, U.S. Marine Corps

Water is a compound made up of two elements—hydrogen and oxygen. The characteristic properties of these elements are different from those of water. However, hydrogen and oxygen have some common properties. They are both colorless, odorless gases, and they both readily react with other elements, making them "reactive" elements. But in many ways they are very different from each other.

Hydrogen has the lowest density of all the elements. It is very reactive, which is one reason why it is present in only very small quantities in air. It reacts with oxygen. You reacted it with oxygen when it burned with a squeaky pop. What do you think was made in that chemical reaction?

It may come as a surprise to you to discover that hydrogen is the most common element in the universe. The sun and other stars are mainly hydrogen gas. Hydrogen is found in many compounds. For example, all acids contain hydrogen.

Oxygen reacts with other substances. Oxygen is needed for burning to take place. Things burn well in oxygen, producing hotter flames. For example, what happened to the glowing splint when it was put into a tube of almost pure oxygen? Some welding and metal-cutting equipment use flammable gases and pure oxygen to produce the high temperatures needed to melt metal.

Oxygen also reacts slowly with many substances. Many compounds containing oxygen are called oxides. You may be familiar with two oxides that are gases—carbon dioxide and sulfur dioxide—but most oxides are solids. In fact, oxygen is the most common element in the Earth's crust, but most of it is combined with other elements to form minerals that make up rocks. ■

▶ WATER IS A COMPOUND FORMED WHEN INFLAMMABLE HYDROGEN REACTS WITH OXYGEN. HERE IT IS BEING USED TO PUT OUT A FIRE. LIKE ALL COMPOUNDS, THE PROPERTIES OF WATER ARE VERY DIFFERENT THAN THOSE OF THE ELEMENTS FROM WHICH IT IS COMPOSED.

PHOTO: U.S. Air Force photo by Senior Master Sgt. David H. Lipp

READING SELECTION

EXTENDING YOUR KNOWLEDGE

CAR BATTERY OR CHEMICAL FACTORY?

Have you ever wondered where the electricity in batteries comes from? How is it possible to store electricity? If you look inside a battery, you can't see electricity! Perhaps the inquiry on the electrolysis of water holds a clue to answering these questions.

In the experiment on the electrolysis of water, you used electrical energy to produce a chemical reaction—splitting the water into hydrogen and oxygen. A battery does the reverse: It uses a chemical reaction to produce an electric current. A battery stores energy as chemical energy.

When a battery is connected to a complete electrical circuit, it releases its stored energy in the form of electricity.

A car battery is not part of a complete electrical circuit until a key is turned in the ignition. Once the key is turned, the electrical circuit is complete, causing a chemical reaction to take place in the car battery. Sulfuric acid and plates of lead metal and lead oxide react to form lead sulfate. During this process, electricity is produced.

▶ CAR BATTERIES STORE ENERGY. DURING A CHEMICAL REACTION, THIS CHEMICAL ENERGY IS TRANSFORMED INTO ELECTRICAL ENERGY.

PHOTO: © David Marsland

Car batteries rarely die, because once the car is started, a generator attached to the engine continually recharges the battery. During recharging, electricity flows through the battery in the opposite direction, reversing the chemical reaction. Car batteries can be recharged thousands of times before they need to be replaced. Many appliances, such as cell phones and cameras, are now manufactured to include rechargeable batteries. ■

DISCUSSION QUESTIONS

1. The inside of a car battery is like a chemical factory. What are the reactants (inputs) and products (outputs) of this chemical reaction?

2. The electric circuit for a car battery is completed upon the turn of the key in the ignition. How are some other electronic circuits completed to make things work?

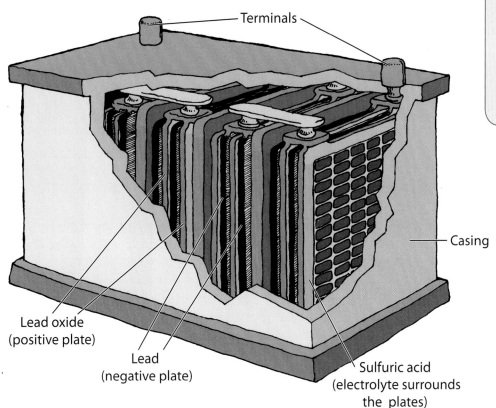

Terminals

Casing

Lead oxide
(positive plate)

Lead
(negative plate)

Sulfuric acid
(electrolyte surrounds
the plates)

▶ THE INSIDE OF A CAR BATTERY IS LIKE A CHEMICAL FACTORY. CHEMICAL ENERGY IS STORED AND THEN RELEASED AS ELECTRICAL ENERGY. WHAT ARE THE REACTANTS AND PRODUCTS OF THIS CHEMICAL REACTION?

THE PROPERTIES OF HYDROGEN
AND THE

DEATH OF AN AIRSHIP

DATE: THURSDAY, MAY 6, 1937

TIME: 7:25 P.M.

LOCATION: LAKEHURST AIR STATION, NEW JERSEY

As the 245-meter-long (804-foot-long) hydrogen-filled airship *Hindenburg* passed over Lakehurst Air Station, it turned to make its final approach. The crew opened ballast tanks to slow its descent, and water gushed from the bottom of the ship, soaking the mooring party below. Everyone at the station was watching the *Hindenburg* as this masterpiece of modern technology hovered about 18 meters (60 feet) above the ground. Near the giant airship, some members of the press adjusted their cameras and sound recorders while others scribbled in their notebooks. In the parking lot, more photographers stood on the tops of cars, attempting to get a better view.

Inside the cabin of the *Hindenburg*, some of the passengers were looking down at the crowds below, searching for the faces of family and friends. They could see the landing crew preparing to catch the ropes dropped from the airship. Suddenly, a deep thump emanated from the stern of the airship. People on the ground began to scream, and the men waiting below the airship began to run. The sky lit up.

Inside the airship, all was chaos. In the officers' mess hall, Werner Franz, a 14-year-old cabin boy, was clearing away plates. As he reached into a cupboard, he felt the whole ship jerk. Plates from the cupboard fell on top of him. He managed to get up and stumble out into the gangway. Everything seemed to be on fire. A huge wall of flames was coming straight at him.

What Werner didn't know was that during docking, the fabric surrounding the hydrogen envelope that kept the ship aloft had somehow ignited. Several theories were presented to explain the fire. Not only did the fabric burn like dry paper, it started an explosive chemical reaction between the hydrogen inside the envelope and the oxygen in the air. The fire became so hot that even the aluminum frame of the airship began to burn. The German engineers who had designed the *Hindenburg* had chosen the wrong materials to make an airship. They had built a floating bomb!

▶ THE HYDROGEN-FILLED AIRSHIP, *HINDENBURG*, FLIES OVER NEW YORK CITY. WHY WAS HYDROGEN USED TO FILL THIS SHIP?

PHOTO: National Air and Space Museum, Smithsonian Institution (SI 92-3598)

Frantically, Werner scrambled away from the flames, toward the front of the airship. The ship lurched again, tilting backward toward the stern. Werner fell and began to slide into the fire. Gathering all his strength, he desperately began to crawl along the floor away from the fire. He could feel the heat through the soles of his shoes. The flames were licking at his legs.

READING SELECTION

EXTENDING YOUR KNOWLEDGE

▶ JUST BEFORE DOCKING, THE *HINDENBURG*
BURST INTO FLAMES.

PHOTO: National Air and Space Museum, Smithsonian
Institution (SI 77-15140)

▶ **THE FIRE CONSUMED THE ENTIRE AIRSHIP. WHAT SUBSTANCE WAS FORMED IN THIS CHEMICAL REACTION?**

PHOTO: National Air and Space Museum, Smithsonian Institution (SI 76-3577)

READING SELECTION
EXTENDING YOUR KNOWLEDGE

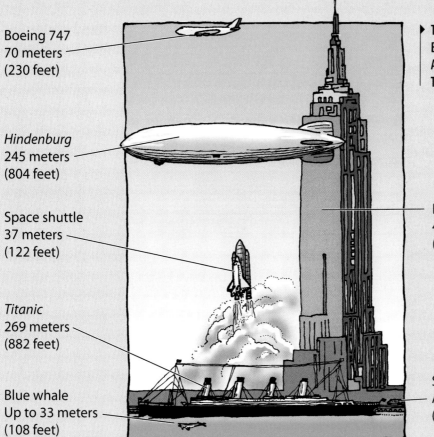

Boeing 747
70 meters
(230 feet)

Hindenburg
245 meters
(804 feet)

Space shuttle
37 meters
(122 feet)

Titanic
269 meters
(882 feet)

Blue whale
Up to 33 meters
(108 feet)

▶ THE *HINDENBURG* WAS
BIGGER THAN A JUMBO JET
AND ALMOST AS LONG AS
THE *TITANIC*.

Empire State Building
443 meters
(1454 feet)

School bus
About 12 meters
(40 feet)

Suddenly, a gush of water knocked Werner flat against the floor. One of the ship's water tanks had burst above him, temporarily extinguishing the nearby fire. But a few seconds later, the fire was back. Werner grabbed at a nearby hatch, kicked it open, and jumped. Winded, he lay on the ground. Screams were still coming from all around. Pulling himself to his feet, he began to run away from the flames. He saw the *Hindenburg*'s captain running in the opposite direction, back to the ship. He was trying to save some of the passengers. Werner turned to run back to help him. As he did so, he was grabbed from behind by an American naval officer, who pulled him to safety.

Thirty-five passengers and crew and one person on the ground died in the flames and wreckage of the *Hindenburg*, as did the dreams of its designers and travel by airship. All of this happened because two elements (which form water!) reacted together to create a disaster. ■

DISCUSSION QUESTIONS

1. How are hot air balloons and airships the same? How are they different?

2. You have probably seen a modern airship, or blimp, on television or over a sports stadium. Use library and Internet resources to find out how these airships work. How are they different from the *Hindenburg*?

Extracting ALUMINUM

PRECIOUS METAL

No one knew about aluminum until 1825. That's when a Danish chemist first extracted pinhead-sized bits of aluminum from a mineral called alumina. But, extracting aluminum from alumina was very difficult, and for most of the 1800s, aluminum was rare and expensive. It was so valuable that kings and queens had fine tea sets and ornamental objects made of aluminum.

▶ THIS ELABORATE LADIES' FAN WITH IMAGES OF THE LINCOLN ASSASSINATION IS MADE OF ALUMINUM.

PHOTO: National Museum of American History, Smithsonian Institution (SI 99-2991)

COMMON METAL

Even though aluminum was once considered very rare, it is the most common metal on Earth, making up 8 percent of Earth's crust. This aluminum is not found as pure aluminum metal but is combined with other elements in the form of aluminum compounds. Today, aluminum is used for everything from airplane frames to soda cans and baseball bats. It is shiny, strong, and lightweight. It doesn't rust and can be shaped and cast. It's even inexpensive enough to use for wrapping leftovers. But aluminum did not become economical until a young inventor working in his backyard lab came up with a way to extract it from alumina.

▶ ALUMINUM IS AN EXCELLENT MATERIAL FOR USE IN LIGHTWEIGHT SPORTS EQUIPMENT, SUCH AS THIS BASEBALL BAT. WHAT OTHER SPORTS EQUIPMENT IS MADE FROM ALUMINUM?

PHOTO: Saquan Stimpson/creativecommons.org

READING SELECTION
EXTENDING YOUR KNOWLEDGE

EARLY START

An eager experimenter, Charles Martin Hall began work on aluminum in 1880. Just 20 years old, he was in his first year at Oberlin College in Ohio.

BACKYARD INVENTOR

Working in a woodshed behind his house, Hall set out to find a way to use electric current to get aluminum metal out of alumina, which contains aluminum and oxygen. The hard part was finding the right liquid in which to dissolve the mineral. Water wouldn't work. Passing electricity through a water solution of alumina only causes the water to break down into hydrogen and oxygen gas. Instead, Hall dissolved it in another mineral, called cryolite. This was tricky. First he had to melt the cryolite by heating it to more than 1000°C (1832°F). He used carbon electrodes to carry the current, because metal electrodes would have melted.

▶ CHARLES MARTIN HALL (1863–1914) DEVELOPED A COMMERCIAL PROCESS FOR EXTRACTING ALUMINUM.

PHOTO: The reproduction of this image is through the courtesy of Alcoa Inc.

▶ HALL'S ORIGINAL SAMPLES OF ALUMINUM REST OF TOP OF HIS NOTES ABOUT THE PROCESS HE INVENTED.

PHOTO: The reproduction of this image is through the courtesy of Alcoa Inc.

▶ IN 1884, BECAUSE IT WAS SUCH A RARE METAL, ALUMINUM WAS GIVEN A PLACE OF HONOR AS A 15-CM HIGH CAP AT THE TOP OF THE WASHINGTON MONUMENT.

PHOTO: The reproduction of this image is through the courtesy of Alcoa Inc.

SUCCESS!

On February 23, 1886, Hall had his first success. After running current through his setup for a few hours, he found several small globs of aluminum inside. Coincidentally, a French scientist named Paul Heroult discovered the same process for obtaining aluminum during the same year. As is often the case with technological innovations, both people had been working to solve the problem of aluminum extraction at the same time. The solution they came up with independently is now dubbed the Hall-Heroult process.

In the U.S., Hall went on to start Alcoa Corporation, still one of the world's largest producers of aluminum. When Hall died in 1914, much of his fortune went to schools around the world. Oberlin College honors his generosity with an aluminum statue. ■

▶ THE MAN AND HIS METAL: THIS STATUE OF CHARLES MARTIN HALL IS MADE FROM ALUMINUM.

PHOTO: The reproduction of this image is though the courtesy of Alcoa Inc.

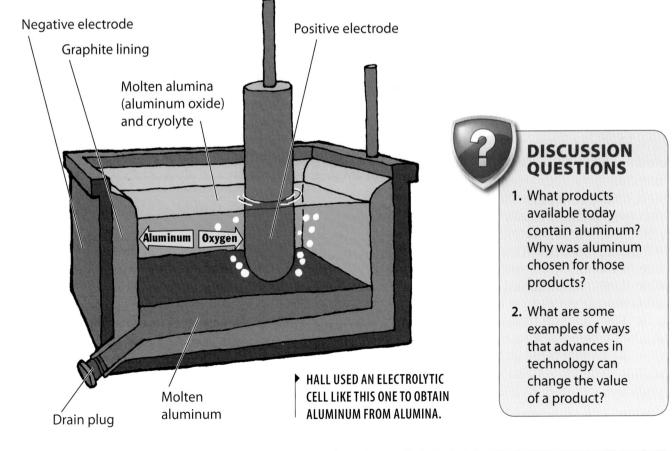

Negative electrode

Graphite lining

Molten alumina (aluminum oxide) and cryolyte

Positive electrode

Aluminum   Oxygen

Molten aluminum

Drain plug

▶ HALL USED AN ELECTROLYTIC CELL LIKE THIS ONE TO OBTAIN ALUMINUM FROM ALUMINA.

DISCUSSION QUESTIONS

1. What products available today contain aluminum? Why was aluminum chosen for those products?

2. What are some examples of ways that advances in technology can change the value of a product?

EXAMINING AND GROUPING ELEMENTS

MARIE CURIE, WORKING WITH HER HUSBAND PIERRE, DISCOVERED TWO ELEMENTS IN 1898 WHILE INVESTIGATING THE RADIOACTIVE ELEMENT URANIUM. THEY NAMED THE FIRST ELEMENT POLONIUM AFTER HER HOME COUNTRY, POLAND. THEY NAMED THE SECOND ELEMENT RADIUM BECAUSE IT WAS VERY RADIOACTIVE. IN 1903, MARIE WAS AWARDED THE NOBEL PRIZE FOR HER WORK. SHE WAS ALSO AWARDED A NOBEL PRIZE IN 1911 FOR HER WORK ON RADIOACTIVE ELEMENTS. MARIE CURIE WAS THE FIRST PERSON TO RECEIVE TWO NOBEL PRIZES.

PHOTO: Library of Congress, Prints & Photographs Division, LC-USZ62-91224

INTRODUCTION

More than 100 different elements exist. They make up all matter. You probably can identify some other elements, such as gold, silver, and aluminum. Can you identify silicon? What about calcium, which is found combined with other elements as a compound in bones and teeth? Did you know that when you breathe, you inhale the elements argon and neon? What are their characteristic properties?

Identifying the elements took scientists hundreds of years. Most elements have been recognized only during the past 60 years. One reason chemists had to work so hard to identify all the elements known today is that most elements are reactive. They combine with other elements to form compounds. You observed a chemical reaction in Lesson 6 when you heard hydrogen burn with a squeaky pop, although you may not have noticed that water vapor was being produced! When you have a large collection of different items, it is useful to put them into groups. You probably have a kitchen drawer containing silverware, divided into knives, forks, and spoons. Items are often classified according to their use and sometimes according to their appearance. Classifying elements has been very helpful to scientists. In this lesson, you will try your hand at classifying some elements.

OBJECTIVES FOR THIS LESSON

Describe the appearance of several elements.

Perform tests and make observations to determine some physical properties of elements.

Collect information on elements and organize it into a table.

Use the information collected to classify elements.

Compare your classification system with one used by chemists.

MATERIALS FOR LESSON 7

For you

1   copy of Student Sheet 7.1a: Examining and Grouping Elements

1   copy of Student Sheet 7.1b: The Periodic Table

For your group

1   black marker

1   sheet of newsprint
     Masking tape

GETTING STARTED

1 Your teacher will refer to Lesson 6 while reviewing the terms "element" and "compound."

2 Participate in a brainstorming session on elements and their characteristic properties.

3 Your teacher will construct a list of your ideas about elements and their characteristic properties. At the end of the lesson, you will look at this list again to discover how much you have learned about elements.

INQUIRY  7.1

INVESTIGATING AND CLASSIFYING ELEMENTS

PROCEDURE

1 Look carefully at Table 1 on Student Sheet 7.1a: Examining and Grouping Elements. You are going to use this table to collect data on 25 different elements. Your teacher will demonstrate how to collect information and will help you complete the table for the elements shown in Figures 7.1 and 7.2. (Use Figure 7.3 when your teacher instructs you to do so.)

▶ ALL THE STARS IN THIS PICTURE OF THE GLOBULAR STAR CLUSTER OMEGA CENTAURI ARE MADE OF A FEW ELEMENTS. WHAT ARE ELEMENTS AND HOW DO THEY DIFFER FROM ONE ANOTHER?

PHOTO: NASA, ESA, and the Hubble SM4 ERO Team

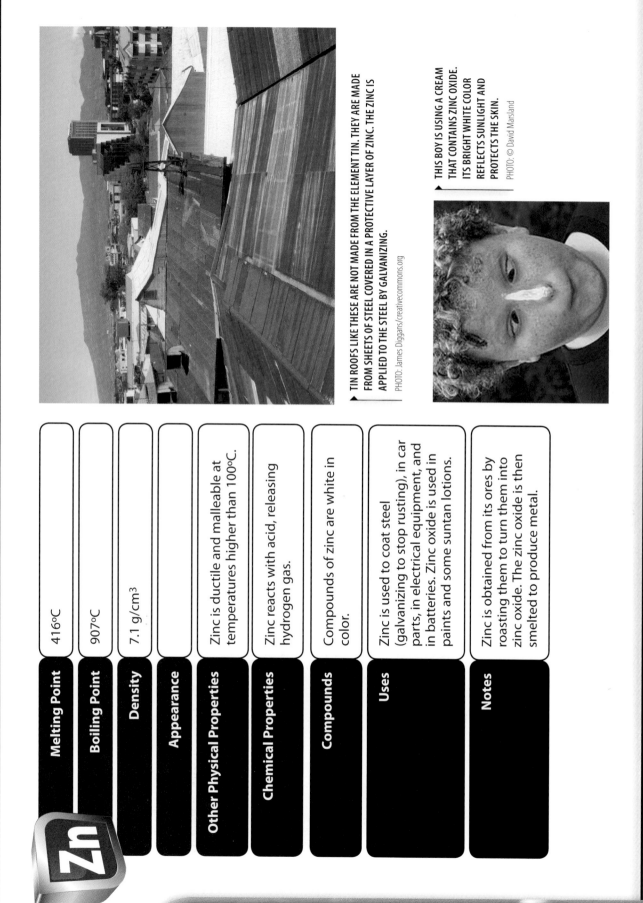

Zn

Melting Point	416°C
Boiling Point	907°C
Density	7.1 g/cm³
Appearance	
Other Physical Properties	Zinc is ductile and malleable at temperatures higher than 100°C.
Chemical Properties	Zinc reacts with acid, releasing hydrogen gas.
Compounds	Compounds of zinc are white in color.
Uses	Zinc is used to coat steel (galvanizing to stop rusting), in car parts, in electrical equipment, and in batteries. Zinc oxide is used in paints and some suntan lotions.
Notes	Zinc is obtained from its ores by roasting them to turn them into zinc oxide. The zinc oxide is then smelted to produce metal.

▶ TIN ROOFS LIKE THESE ARE NOT MADE FROM THE ELEMENT TIN. THEY ARE MADE FROM SHEETS OF STEEL COVERED IN A PROTECTIVE LAYER OF ZINC. THE ZINC IS APPLIED TO THE STEEL BY GALVANIZING.
PHOTO: James Diggans/creativecommons.org

▶ THIS BOY IS USING A CREAM THAT CONTAINS ZINC OXIDE. ITS BRIGHT WHITE COLOR REFLECTS SUNLIGHT AND PROTECTS THE SKIN.
PHOTO: © David Marsland

FIGURE 7.1 USE THE FACTS AND PHOTOS SHOWN HERE AND THE TESTS DEMONSTRATED BY YOUR TEACHER TO COMPLETE THE INFORMATION FOR ZINC IN TABLE 1 ON STUDENT SHEET 7.1A.

Uranium

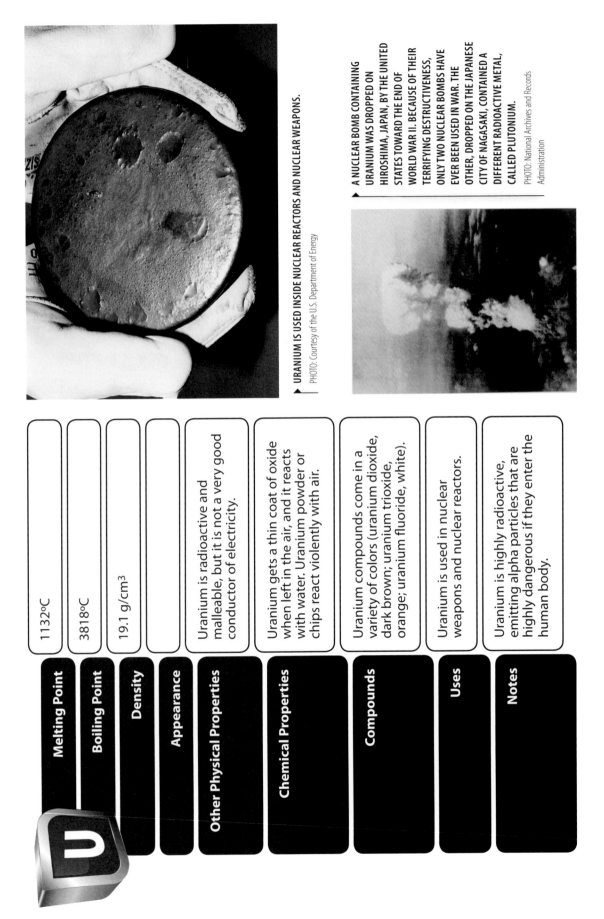

URANIUM IS USED INSIDE NUCLEAR REACTORS AND NUCLEAR WEAPONS.
PHOTO: Courtesy of the U.S. Department of Energy

A NUCLEAR BOMB CONTAINING URANIUM WAS DROPPED ON HIROSHIMA, JAPAN, BY THE UNITED STATES TOWARD THE END OF WORLD WAR II. BECAUSE OF THEIR TERRIFYING DESTRUCTIVENESS, ONLY TWO NUCLEAR BOMBS HAVE EVER BEEN USED IN WAR. THE OTHER, DROPPED ON THE JAPANESE CITY OF NAGASAKI, CONTAINED A DIFFERENT RADIOACTIVE METAL, CALLED PLUTONIUM.
PHOTO: National Archives and Records Administration

Melting Point	1132°C
Boiling Point	3818°C
Density	19.1 g/cm³
Appearance	
Other Physical Properties	Uranium is radioactive and malleable, but it is not a very good conductor of electricity.
Chemical Properties	Uranium gets a thin coat of oxide when left in the air, and it reacts with water. Uranium powder or chips react violently with air.
Compounds	Uranium compounds come in a variety of colors (uranium dioxide, dark brown; uranium trioxide, orange; uranium fluoride, white).
Uses	Uranium is used in nuclear weapons and nuclear reactors.
Notes	Uranium is highly radioactive, emitting alpha particles that are highly dangerous if they enter the human body.

FIGURE 7.2 USE THE FACTS AND PHOTOS SHOWN HERE TO HELP YOU COMPLETE THE INFORMATION FOR URANIUM IN TABLE 1 ON STUDENT SHEET 7.1A.

Helium

He

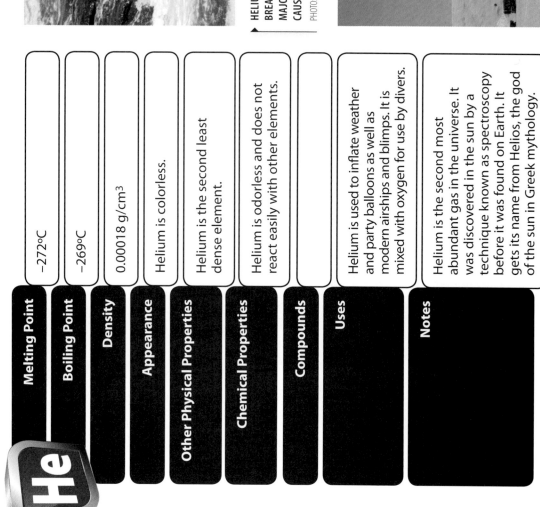

Melting Point	–272°C
Boiling Point	–269°C
Density	0.00018 g/cm³
Appearance	Helium is colorless.
Other Physical Properties	Helium is the second least dense element.
Chemical Properties	Helium is odorless and does not react easily with other elements.
Compounds	
Uses	Helium is used to inflate weather and party balloons as well as modern airships and blimps. It is mixed with oxygen for use by divers.
Notes	Helium is the second most abundant gas in the universe. It was discovered in the sun by a technique known as spectroscopy before it was found on Earth. It gets its name from Helios, the god of the sun in Greek mythology.

◀ HELIUM/OXYGEN MIXTURES ARE USED BY DIVERS WHO DESCEND TO GREAT DEPTHS. WHEN BREATHED UNDER PRESSURE, HELIUM HAS LESS TOXIC EFFECTS THAN NITROGEN, THE MAJOR COMPONENT OF AIR. WHEN NITROGEN IS BREATHED UNDER PRESSURE, IT CAN CAUSE SYMPTOMS OF NITROGEN NARCOSIS, A CONDITION THAT CAN DISORIENT DIVERS.

PHOTO: Roger H. Goun/Work licensed under a Creative Commons Attribution License

▲ ONLY HYDROGEN IS LESS DENSE THAN HELIUM. HELIUM IS USED TO FILL BALLOONS SUCH AS THIS NASA RESEARCH BALLOON. WHY IS HELIUM USED FOR THIS PURPOSE INSTEAD OF HYDROGEN?

PHOTO: NASA

FIGURE **7.3** WHEN INSTRUCTED BY YOUR TEACHER, USE THE FACTS AND PHOTOS SHOWN HERE TO ENTER THE INFORMATION FOR HELIUM IN TABLE 1 ON STUDENT SHEET 7.1A.

Inquiry 7.1 continued

2 Using your observations from Lesson 6 and your own knowledge, complete the rows in Table 1 for hydrogen and oxygen.

3 Participate in a class discussion of your information on hydrogen and oxygen in Table 1. If additional information about these elements arises during the discussion, add it to your table.

4 Your teacher will explain how you should investigate some other elements. Follow along as your teacher reviews Steps 5 through 9 of this procedure.

5 The elements are arranged in stations around the room. For each element, there is a card that looks similar to Figure 7.1, 7.2, and 7.3. Some of these cards have missing information. You will need to examine and test the elements to determine what some of their properties are.

6 For some elements, you must determine whether they conduct electricity (allow electricity to pass through them) or whether they are insulators (do not allow electricity to pass through them). See Figure 7.4 for information on how to perform this test.

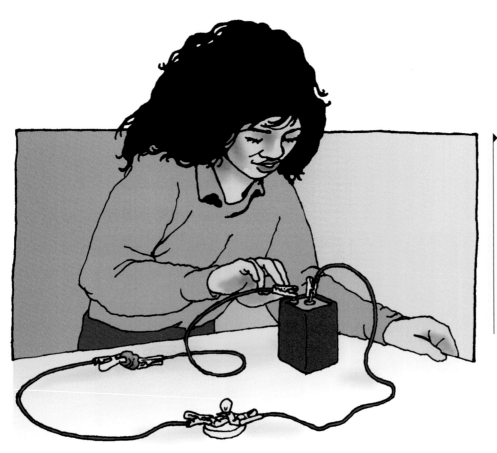

▶ FOR SOME ELEMENTS, YOU WILL PERFORM A TEST TO DETERMINE WHETHER THEY CONDUCT ELECTRICITY. MAKE A CONDUCTIVITY-TESTING APPARATUS. TEST THE ELEMENTS AS SHOWN. IF THE BULB LIGHTS, THE ELEMENT CONDUCTS ELECTRICITY. THE BRIGHTER THE BULB, THE BETTER THE ELEMENT CONDUCTS ELECTRICITY.
FIGURE **7.4**

7 If there is a paper clip at the station, you should investigate the hardness of the element (see Figure 7.5). First, try scratching the element with your fingernail. If this has no effect, try using the end of the paper clip. Is the element hard or soft compared with your fingernail and the paper clip? What does the scratched surface of the element look like?

8 Use the magnet to determine whether an element is magnetic.

9 You and your partner will be assigned to a numbered station. Go to that station and start collecting the information you need to complete Table 1. Do not transfer all of the information on the card to Table 1. Select only the information that you think you need. (Remember that the photographs and their captions contain useful information.) You have 5 minutes to investigate each element.

10 When your teacher calls time, leave the card, element, and any apparatus where you found it. Move to the next station. (If you are at Station 6, move to Station 1.)

▶ USE YOUR FINGERNAIL AND THE PAPER CLIP TO DETERMINE HOW HARD THE ELEMENT IS. ALSO OBSERVE HOW THE SURFACE OF THE ELEMENT LOOKS AFTER IT HAS BEEN SCRATCHED.
FIGURE **7.5**

Inquiry 7.1 continued

11 When you have collected information on all of the elements at the stations, return to your seat. Your teacher will outline how you should place your elements in groups and will ask for your ideas on grouping the elements.

12 Working with the other pair in your group, try to identify at least five groups of elements. Place the elements in these groups. Remember that most elements will fit into more than one group.

13 Write your ideas in your science notebook. When you think you have useful groups of elements, transfer the information from your notebook to the newsprint (see Figure 7.6). When you have finished writing all five groups, attach the sheet of newsprint to the wall.

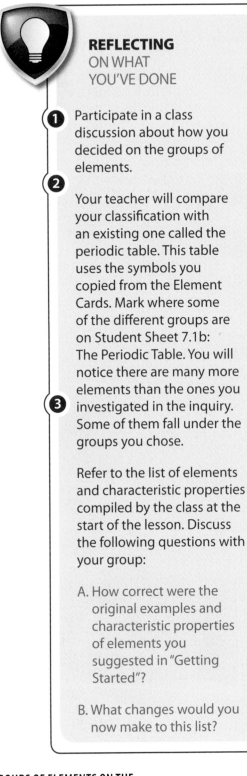

REFLECTING ON WHAT YOU'VE DONE

1 Participate in a class discussion about how you decided on the groups of elements.

2 Your teacher will compare your classification with an existing one called the periodic table. This table uses the symbols you copied from the Element Cards. Mark where some of the different groups are on Student Sheet 7.1b: The Periodic Table. You will notice there are many more elements than the ones you investigated in the inquiry. Some of them fall under the groups you chose.

3 Refer to the list of elements and characteristic properties compiled by the class at the start of the lesson. Discuss the following questions with your group:

A. How correct were the original examples and characteristic properties of elements you suggested in "Getting Started"?

B. What changes would you now make to this list?

▸ RECORD YOUR GROUPS OF ELEMENTS ON THE NEWSPRINT. MAKE SURE THE LETTERING IS EASY TO READ FROM THE OTHER SIDE OF YOUR CLASSROOM.
FIGURE **7.6**

DMITRY'S Card Game

I t's interesting how different skills can be brought together to contribute to scientific discovery. Consider the story of Dmitry Mendeleyev and his card game that changed the face of chemistry.

Mendeleyev was a Russian college professor who loved to play cards. He was also looking for a way to organize the elements. It is thought that Mendeleyev may have used his love of cards to help accomplish this feat. As the story goes, he first wrote the symbols, characteristic properties, and other information for 63 elements on cards (only 63 elements had been discovered by 1869, the year he developed the card game). He then placed the cards face up on a table and began moving them around. He placed them in order of the mass of their atoms, or atomic mass. Then he compared their other properties. He discovered that when he did this, he could see patterns emerging. He noticed that certain properties periodically repeated themselves at the same interval. He began to group the cards with similar properties together in rows and columns. For example, the elements sodium and potassium, which are next to each other in his original table, are soft, shiny, and highly reactive metals. He continued to group elements with similar properties together until he had established a complete table of all 63 elements. Because his table was based on the pattern of properties repeating themselves, he called this classification system the "Periodic Table of the Elements."

▶ MENDELEYEV (1834–1907) WAS 35 YEARS OLD WHEN HE PUBLISHED HIS "PERIODIC TABLE OF THE ELEMENTS." THIS CHARCOAL DRAWING, DONE BY HIS WIFE, SHOWS HIM IN HIS LATER YEARS, WORKING IN HIS LABORATORY. PART OF HIS PERIODIC TABLE CAN BE SEEN IN THE BACKGROUND OF THIS PICTURE.

PHOTO: Courtesy of Smithsonian Institution Libraries, Washington, D.C.

▶ MENDELEYEV'S PERIODIC TABLE ADORNS THE WALL OF THE SCHOOL WHERE HE WORKED IN ST. PETERSBURG, RUSSIA. THIS TABLE INCLUDES THE NOBLE GASES (HE, NE, AR) DISCOVERED AFTER HIS ORIGINAL TABLE OF 63 ELEMENTS.

PHOTO: dushenka/creativecommons.org

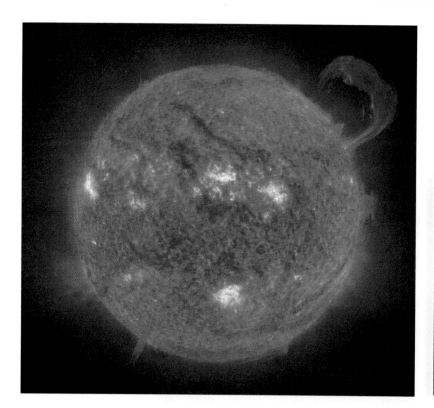

THE PERIODIC TABLE IS THE "LIST OF INGREDIENTS" FOR OUR ENTIRE UNIVERSE. MOST OF THE ELEMENTS WERE FORMED IN NUCLEAR REACTIONS, WHICH TOOK PLACE INSIDE STARS OR EXPLODING STARS CALLED SUPER NOVA. SOME OF THESE NUCLEAR REACTIONS CONTINUE TO TAKE PLACE INSIDE OUR SUN (SHOWN IN THIS IMAGE), WHICH RELEASES ENERGY WHEN THE ELEMENT HYDROGEN UNDERGOES FUSION AND IS CONVERTED TO THE ELEMENT HELIUM.

PHOTO: ESA/NASA/SOHO

Over a period of six years, Mendeleyev improved his classification system. There were some gaps in the table—missing cards. He predicted that elements yet to be discovered would fill these gaps. He was able to suggest some of the characteristic properties they would have. As scientists became more knowledgeable about the physics and chemistry of matter, they helped refine the table, and the missing elements were discovered. ■

DISCUSSION QUESTIONS

1. The periodic table was not the work of one person. Evidence was collected over many years by scientists in many countries. Use the Internet and other resources to find out about the scientists involved in developing the periodic table.

2. Mendeleyev created his periodic table by grouping elements according to their properties. In what other ways do we use classification in science?

COMBINING ELEMENTS

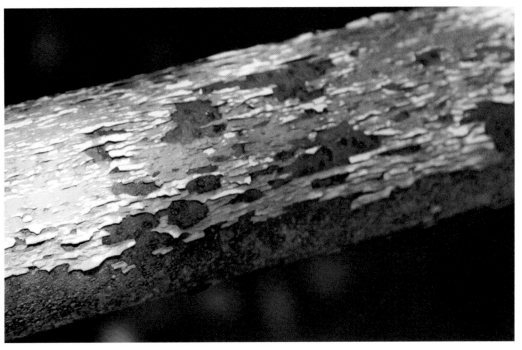

▶ WHAT IS GOING ON HERE? IS A CHEMICAL REACTION TAKING PLACE? WHAT ARE THE REACTANTS AND THE PRODUCTS OF THIS REACTION?

PHOTO: National Science Resources Center

INTRODUCTION

In the previous lesson, you discovered that elements can be classified into groups based on their characteristic properties. In Inquiry 8.1, you will return to this theme and attempt to classify elements into two major groups. Your classification will be used as a springboard to investigate the chemical properties of one of these groups in more detail. In Inquiry 8.2, you will investigate how two elements from the two groups you have identified react to make a compound. You will compare some of the properties of the reactants and products of the reaction and use a simple word equation to describe the reaction that has taken place.

OBJECTIVES FOR THIS LESSON

Examine the properties of four elements.

Place the four elements into two major groups.

Identify these groups on the periodic table.

Make a compound from elements in these two groups.

Construct a simple word equation for the reaction that has taken place.

Discuss the differences between reactants and products.

▶ **MATERIALS FOR LESSON 8**

For you

    Your copy of Student Sheet 7.1a: Examining and Grouping Elements

    Your copy of Student Sheet 7.1b: The Periodic Table

  1   copy of Student Sheet 8.1: Splitting the Periodic Table

  1   copy of Student Sheet 8.2: Reacting Two Elements

  1   pair of safety goggles

For you and your lab partner

  2   test tubes

  1   250-mL beaker

  1   pair of scissors

  1   index card

  1   piece of steel wool

  1   metric ruler

     Masking tape

For your group

  1   bolt

  1   cylinder

  1   lump of yellow solid

  1   lump of black solid

     Access to a wall clock

GETTING STARTED

1 In Lesson 7, you grouped elements according to their properties. Your teacher will review some of these groups. Be prepared to contribute the name of a group you identified.

2 Imagine you have to divide the elements you investigated into two groups. Discuss the following questions with your group:

A. How would you select the groups?

B. What properties (criteria) would you use to decide which elements go into which groups? (The best groupings may use more than one property as criteria.)

C. What names would you give each group?

D. What are some elements you would put into each group?

3 Write your ideas in your science notebook and share them in the class discussion. ☞

> **! SAFETY TIP**
> Wear your safety goggles at all times.

INQUIRY 8.1

SPLITTING THE PERIODIC TABLE

PROCEDURE

1 One student from your group should pick up a plastic box containing the materials.

2 Check the contents of the plastic box against the materials list. (The plastic box does not contain the piece of steel wool. You will get that later.)

3 Take out the bolt, the cylinder, and the yellow and black samples. These are all elements. Examine them closely. Your teacher will ask you to identify them.

4 Sort these elements into the two groups you identified in "Getting Started." If they do not fit easily into the two groups, try changing the criteria you used to select the groups. If necessary, decide on a new name for each group. Construct a table in your science notebook that compares the properties of the elements in the two groups. Compare as many different properties as you can. Use the information you collected about these elements in Student Sheet 7.1a (Table 1) to help you. ☞

REACTING TWO ELEMENTS

5 Your teacher will discuss the groups you selected and choose two of the groups to conduct a class brainstorming session. Your teacher will record the properties of each group on a Venn diagram. At the end of the brainstorming session, copy the completed Venn diagram into your notebook.

6 Label the Venn diagram on Student Sheet 8.1: Splitting the Periodic Table with the names of the two groups. Try to place the elements you have encountered (both in the lessons and in your own experience) into one of the groups. If any elements seem to have intermediate properties (properties between both groups), place them in the area where the circles overlap.

7 Your teacher will ask you to give the names of elements and where you placed them on the Venn diagram.

8 Look at the periodic table on Student Sheet 7.1b. Is it possible to draw a line through the table that separates the nonmetals from the metals? Discuss this with your group and then use a pencil to draw the line.

PROCEDURE

1 In this inquiry you will study the chemical properties of metals in more detail, starting with the reaction between a metal and a nonmetal. Your teacher will begin by reviewing the procedure you need to follow.

2 If the index card does not already have holes in it, cut two round holes, each with a diameter that is slightly larger than the diameter of the test tubes (see Figure 8.1).

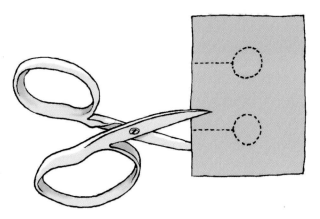

▶ CUT TWO HOLES IN THE INDEX CARD. EACH HOLE SHOULD HAVE A DIAMETER SLIGHTLY LARGER THAN THE DIAMETER OF THE TEST TUBES.
FIGURE **8.1**

Inquiry 8.2 continued

3 Put 150 mL of water into the beaker. Use masking tape to attach the index card to the top of the beaker (see Figure 8.2).

▸ USE TAPE TO AFFIX THE INDEX CARD TO THE TOP OF THE BEAKER.
FIGURE **8.2**

4 Invert an empty test tube. Place the test tube through one of the holes in the card, making sure the open end of the test tube is resting on the bottom of the beaker (see Figure 8.3). You may need to tape the tube in place.

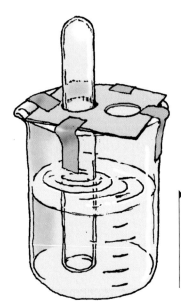

▸ PLACE AN EMPTY TEST TUBE THROUGH THE INDEX CARD WITH THE OPEN END OF THE TEST TUBE RESTING ON THE BOTTOM OF THE BEAKER.
FIGURE **8.3**

5 In Table 1 on Student Sheet 8.2: Reacting Two Elements, fill in the diagram of the empty test tube to show the water level in the test tube.

6 Pick up a piece of damp steel wool from your teacher. The steel wool has been dipped in vinegar to clean off any grease or dirt. Steel wool is mainly iron with a little carbon added. For the purposes of this inquiry, it can be treated as if it is pure iron.

7 Quickly put the steel wool into the second test tube. Push it down to the bottom of the tube with a pencil.

8 Place the test tube through the other hole in the card, making sure the open end of the test tube is resting on the bottom of the beaker (see Figure 8.4).

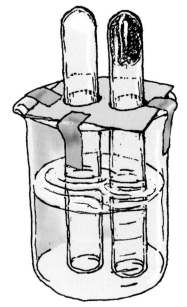

▸ THE COMPLETED APPARATUS FOR INQUIRY 8.2.
FIGURE **8.4**

9 Immediately show the water level by filling in the diagram in Table 1. Write a description of the damp steel wool in the third column of the table.

10 Watch the apparatus carefully. Do you notice anything happening in the tubes?

11 After about 15 minutes, look at the level of the water in each tube. Record your observations in the appropriate places in Table 1.

12 Describe the appearance of the steel wool.

13 Air is about 6 percent oxygen. Keeping the tubes in position, remove the card and use the ruler to measure how far up the test tube the water has moved. Answer these questions on Student Sheet 8.2:

• Has the water level changed in either tube? If so, can you explain the change?

• What can you conclude from your observations?

14 Clean the test tubes and the beaker. Make sure you remove the steel wool from the test tube and place it in the trash. Return the cleaned apparatus and the index card to the plastic box.

REFLECTING
ON WHAT
YOU'VE DONE

1 Discuss your results and compare them with those of other pairs.

2 Answer the questions in Steps 4–7 on Student Sheet 8.2.

All chemical reactions have reactants (inputs) and products (outputs) and can be written as simple word equations. For example, you know that hydrogen combines with oxygen to form water. A simple word equation to describe this equation is as follows:

hydrogen + oxygen = water

3 On Student Sheet 8.2, complete Step 8 by writing a word equation for the reaction that took place in the test tube. Label the reactants and the products in the equation. This is an example of the synthesis of a new product from a chemical reaction. In Lesson 10, you will investigate how metals react with acids to form new products.

SYNTHESIZING MATERIALS NEW

Imagine what it would be like to invent a new substance. Many of the materials that we take for granted were invented. These materials are synthetic, which means they don't exist in nature. They have been made—or synthesized—from natural or other substances. Can you think of any examples of synthesized substances?

Probably the most well known group of synthetic substances is plastics. Some of the early plastics were made by altering natural substances, such as cellulose and latex, which are found in plants. The first plastic that did not rely on starting materials from nature was called Bakelite™, after its inventor, Leo Baekeland (1863-1944). In 1907, he found a way to control a chemical reaction between two substances to produce a brittle, dark brown plastic that was, because of its insulating properties, used for making electrical fittings and household items.

▶ THIS VINTAGE TELEVISION WAS MADE FROM BAKELITE™, AN EARLY PLASTIC.

PHOTO: Courtesy of www.tvhistory.tv

IN 1931, WALLACE CAROTHERS (1896–1937), WORKING AT DUPONT, INVENTED A SILK-LIKE SYNTHETIC PLASTIC THAT WAS EVENTUALLY CALLED NYLON. HE MADE IT BY MIXING TOGETHER AN ACID AND A SOLUTION OF ANOTHER SUBSTANCE, DIAMINE. NYLON, WHEN USED AS A FIBER, HAS MANY OF THE PROPERTIES OF SILK, BUT IS STRONGER. IN THIS PICTURE, CAROTHERS DEMONSTRATES ANOTHER SYNTHETIC COMPOUND— A TYPE OF RUBBER.

PHOTO: Courtesy of the Hagley Museum and Library

NYLON CAME TO THE RESCUE DURING WORLD WAR II. SILK, PRODUCED BY SILKWORMS, HAD BEEN USED IN THE PAST TO MAKE PARACHUTES BUT WAS IN SHORT SUPPLY.

PHOTO: Jason Gulledge/creativecommons.org

SYNTHETIC MATERIALS HELP WIN A WAR

Many new synthetic plastics were first produced in bulk in the 1930s, just in time to play an important role in World War II. Here are some examples, which contributed to the victory of the Allies.

▶ NYLON WAS USED IN THE MANUFACTURE OF PARACHUTES FOR AIRCREW AND PARATROOPERS.

PHOTO: National Archives and Records Administration

▶ NYLON AND SILK STOCKINGS WERE RECYCLED TO ENSURE ADEQUATE SUPPLIES.

PHOTO: National Archives and Records Administration/
Franklin D. Roosevelt Library

▶ WHEN THE JAPANESE ARMY TOOK OVER THE RUBBER PLANTATIONS IN THE FAR EAST, RUBBER WAS IN SHORT SUPPLY. THE TIRES ON THIS ARMY TRUCK WERE MADE FROM A NEW PLASTIC, SOMETIMES CALLED SYNTHETIC RUBBER.

PHOTO: National Archives and Records Administration/
Franklin D. Roosevelt Library

▶ POLYVINYL CHLORIDE (PVC) WAS
USED TO INSULATE ELECTRICAL WIRES
INSIDE AIRCRAFT.

PHOTO: Library of Congress, Prints & Photographs
Division, FSA-OWI Collection, LC-USW36-273

▶ POLYACRYLICS (FOR EXAMPLE, PLEXIGLASS™) WERE
TRANSPARENT AND LIGHT AND DIDN'T SHATTER
LIKE GLASS—IDEAL PROPERTIES WHEN IT CAME TO
AIRCRAFT MANUFACTURE.

PHOTO: National Archives and Records Administration/Franklin D. Roosevelt Library

Hundreds of synthetic plastics are in
use today, all with different properties.
They are used in products as varied
as soda bottles, lenses, and artificial
body parts. ■

DISCUSSION QUESTIONS

1. What are some examples
of synthetic materials you
encounter in everyday life?

2. What are the pros and cons
of our extensive use of
synthetic materials today?

Alchemy *Into* CHEMISTRY

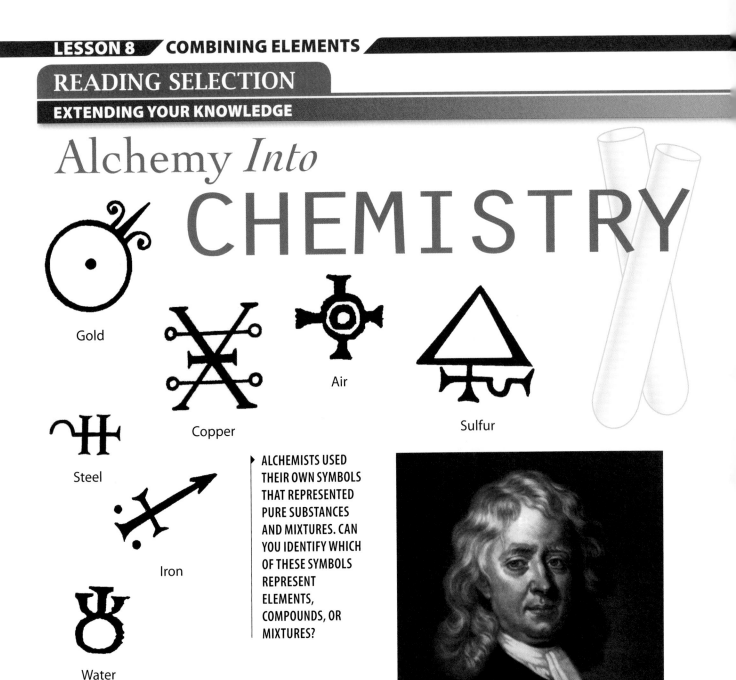

Gold

Copper

Air

Steel

Sulfur

Iron

Water

▶ ALCHEMISTS USED THEIR OWN SYMBOLS THAT REPRESENTED PURE SUBSTANCES AND MIXTURES. CAN YOU IDENTIFY WHICH OF THESE SYMBOLS REPRESENT ELEMENTS, COMPOUNDS, OR MIXTURES?

S ince the dawn of civilization, chemical reactions have seemed magical. Certain rocks weep molten metal if put in a fire. Two substances mixed together burst into flames. Juice from a certain plant cures illness. It is no surprise, then, that early thinkers mixed magic with observation and experiment as they tried to understand the world.

SIR ISAAC NEWTON

▶ SCIENTIST OR ALCHEMIST? ISAAC NEWTON (1642–1727) WAS ONE OF THE GREATEST SCIENTISTS THE WORLD HAS EVER KNOWN. A PHYSICIST AND MATHEMATICIAN, HE LIVED AND WORKED IN THE 1600S, JUST WHEN MODERN SCIENCE WAS TAKING HOLD. NEWTON INVENTED A FORM OF CALCULUS AND DISCOVERED THE LAWS THAT GOVERN THE MOTION OF THE PLANETS. HE WAS ALSO FASCINATED BY AND LEARNED A LOT FROM ALCHEMY.

PHOTO: Library of Congress, Prints & Photographs Division, LC-USZ62-10191

▶ THIS DRAWING SHOWS AN ALCHEMIST'S LAB IN 16TH-CENTURY EUROPE. THE BIG POT IN THE MIDDLE OF THE FURNACE WAS USED FOR PURIFYING LIQUIDS BY THE SAME DISTILLATION PROCESS THAT IS NOW USED FOR THE DESALINIZATION OF SALT WATER.

PHOTO: Library of Congress, Prints & Photographs Division, LC-USZ62-80071

The study of matter in ancient times began as alchemy, a mishmash of primitive chemistry, superstition, and showmanship. Alchemy had two main magical aims: to change common metals into gold and to find a medicine that would cure all ills, including old age. By the 1600s, alchemists were slowly learning that observation and experiment provided more useful information than magic and sorcery. They learned to make hypotheses, gather evidence, and form conclusions. The modern science of chemistry was born. ∎

DISCUSSION QUESTIONS

1. How did the practice of alchemy evolve into the process of chemistry?

2. Distillation involves vaporization and condensation of a liquid to separate its components. Think about how you might make a distillation apparatus from things you have at home, and draw a picture of your idea.

EXPLORATION ACTIVITY: RESEARCHING A COMPOUND

▶ CAN YOU GUESS WHAT SIMPLE COMPOUND WAS USED TO CONSTRUCT THIS TABLE?

PHOTO: Jimmy Harris/
creativecommons.org

INTRODUCTION

In this lesson, you will begin an Exploration Activity that you will work on over the next several weeks. This project gives you the opportunity to apply what you have learned in the unit to the world around you. In this research activity, you and a partner will select a simple compound. You will investigate the chemistry, uses, and history of the compound by doing research at the library and on the Internet. You will compile the information you collect to create an exhibit. You will also give an oral presentation on the information you have found. The work you do for the Exploration Activity will be an important part of your grade for this unit.

OBJECTIVES FOR THIS LESSON

Select a simple compound that is made up of two or three elements to research.

Conduct library and Internet research on the structure, history, occurrence and uses of the compound you have chosen.

Create an exhibit based on your research.

Give an oral presentation on the compound you selected.

MATERIALS FOR LESSON 9

For you

1   copy of Student Sheet 9a: Researching Salt as a Compound

1   copy of Student Sheet 9b: Exploration Activity Schedule

1   copy of Inquiry Master 9a: Scoring Rubric for the Cube

1   copy of Inquiry Master 9b: Scoring Rubric for the Oral Presentation

For you and your partner

Clear tape or glue

Scissors

Card stock, poster board, or lightweight cardboard

GETTING STARTED

1 Read "A Brief History of Salt" on pages 114-119, and answer the accompanying questions on Student Sheet 9a: Researching Salt as a Compound.

2 Contribute to a class discussion on the reading selection.

3 Use the information from the table provided by your teacher to complete the table on Student Sheet 9a.

▶ LIMESTONE IS A COMPOUND. WHAT ELEMENTS COMPRISE LIMESTONE?

PHOTO: andy nunn/ creativecommons.org

INTRODUCING THE EXPLORATION ACTIVITY

PROCEDURE

1 After your teacher gives you Student Sheet 9b: Exploration Activity Schedule, tape it to the inside front cover of your science notebook. You will need to refer to it as you work on the Exploration Activity. Listen carefully as your teacher gives you due dates and point values for each component. Record this information on your schedule. Follow it carefully, or you may lose points.

2 Follow along as your teacher reviews the Exploration Activity guidelines found in this Student Guide. Ask questions about anything that isn't clear to you.

EXPLORATION ACTIVITY GUIDELINES

PART 1
CHOOSING THE COMPOUND
PROCEDURE

1 To make your work easier, you and your partner should choose a relatively simple and well-known (to scientists, anyway) compound. Discuss with your partner which compound you are going to research. You do not need to make a final decision immediately. Some examples are given in the list titled "Exploration Activity Compounds." You can choose one of these or look for one in the resources in your school library or online. But remember: choose a compound that has the basic information available. A familiar compound that readily exists in nature, made from two or three elements, will be much easier to research and present than a rare compound.

EXPLORATION ACTIVITY COMPOUNDS

This is a list of elements, their familiar compounds, and the chemical formulas for those compounds:

- Hydrogen compounds—acids, hydrides, and hydroxides

    Hydrochloric acid (HCl)

    Boric acid (H_3BO_4)

    Carbonic acid (H_2CO_3)

    Sulfuric acid (H_3PO_4)

    Nitric acid (HNO_3)

    Hydrogen peroxide (H_2O_2)

    Methane (CH_4)

    Ammonia (NH_3)

    Butane gas (C_4H_{10})

- Carbon compounds or organic compounds, which make up a separate branch of chemistry

    Carbon dioxide (CO_2), a major greenhouse gas

    Carbon tetrachloride (CCl_4), a dry-cleaning compound

    Freon (CCl_3F) in air conditioners

- Oxygen compounds—oxides

    Silicon dioxide (SiO_2)

    Magnesium oxide (MgO)

    Iron (II) oxide (Fe_2O_3)

    Nitrous oxide (N_2O)

    Sulfur dioxide (SO_2)

- Sodium, magnesium, phosphorus, and calcium compounds—abundant in earth's crust

    Sodium oxide (NaO)

    Sodium hydroxide (NaOH), a drain cleaner

    Sodium bicarbonate ($NaHCO_3$), baking soda

    Sodium thiosulfate (NaS_2O_3), a fixing agent in photo printing

    Magnesium chloride (MgCl), abundant in seawater

    Magnesium oxide (MgO), 35% of earth's crust

    Hydrogen phosphate (H_3PO_4), in soft drinks, fertilizers, and solvents

    Calcium oxide (CaO), lime

    Calcium carbonate ($CaCO_3$), limestone

    Calcium sulfate ($CaSO_4$), in gypsum and plaster of paris

    Calcium phosphate ($CA_3(PO_4)$), the primary inorganic substance in bones and teeth

- Sulfur compounds

    Hydrogen sulfide (H_2S), has a "rotten egg" odor

    Sulfur oxide (SO_2), a refrigerant, bleaching agent, and disinfectant

    Hydrogen sulfate (H_2SO_4), in fertilizer, pigments, and dyes

Exploration Activity Part 1 continued

2 During the next week, meet with your partner and write a short paragraph identifying the compound you have chosen. Give the reasons for your choice. Figure 9.1 shows a sample paragraph. Hand it in by the due date on your schedule. Your teacher must approve your compound. If too many pairs choose the same compound, you may be asked to select another.

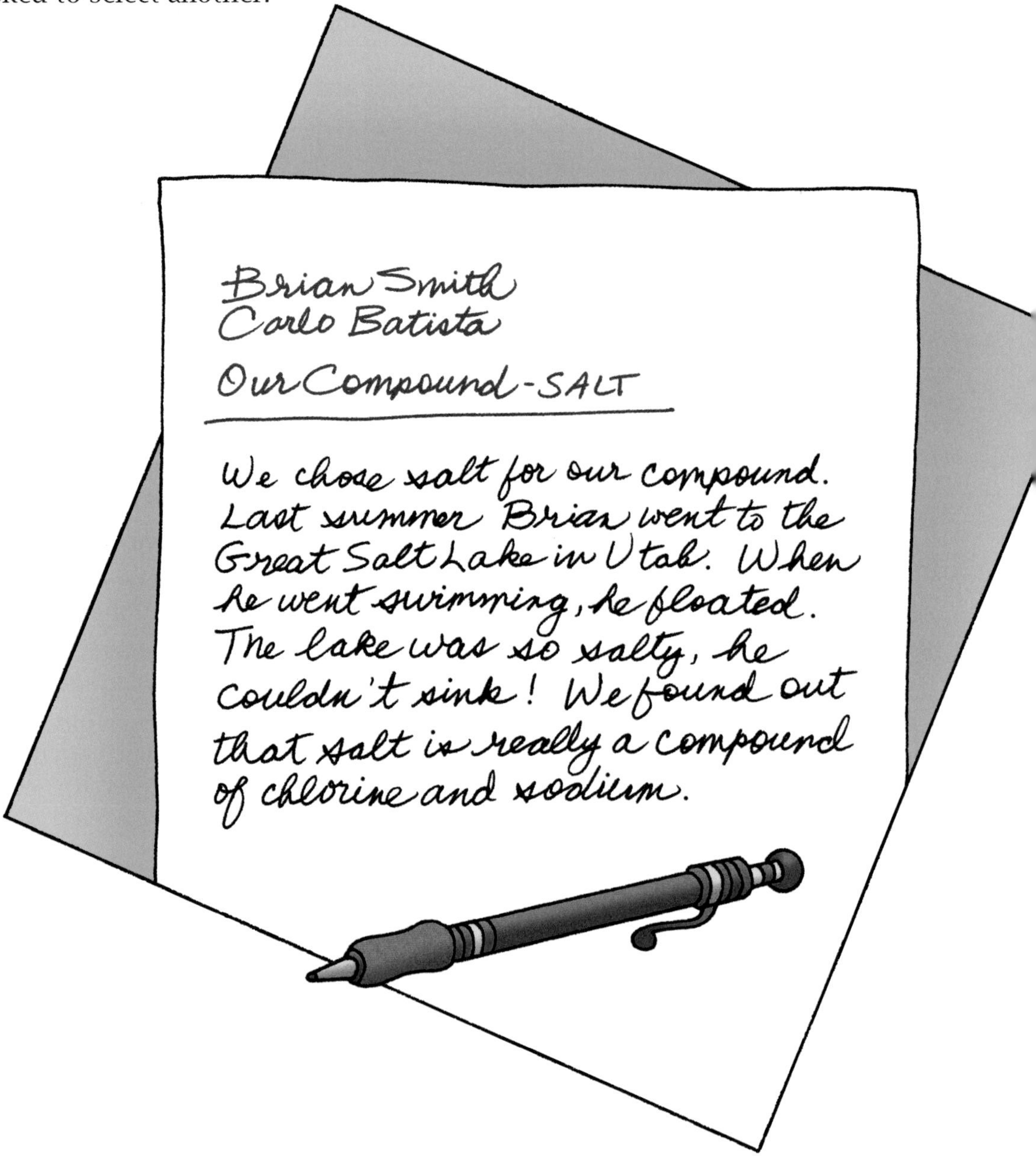

> Brian Smith
> Carlo Batista
> Our Compound - SALT
>
> We chose salt for our compound. Last summer Brian went to the Great Salt Lake in Utah. When he went swimming, he floated. The lake was so salty, he couldn't sink! We found out that salt is really a compound of chlorine and sodium.

▶ SAMPLE PARAGRAPH IDENTIFYING THE COMPOUND AND THE REASON FOR YOUR CHOICE.
FIGURE 9.1

PART 2

STARTING THE RESEARCH

PROCEDURE

1 Start gathering information about your compound. Your information will be divided into five sections. As you gather information, write your notes under these headings:

COMPOUND Explain the properties of the compound. Include the chemical formula that shows the elements in it.

COMBINING ELEMENTS Give the chemical and physical properties of the elements which make up the compound.

USES OF THE COMPOUND Explain how the uses of the compound are related to its properties. Who used it and for what? Has its use changed today?

OCCURRENCE OF THE COMPOUND Explain where it is found in nature or how it is produced in the laboratory.

HISTORY OF THE COMPOUND When and where did it first appear? Who investigated its structure in the laboratory?

2 Use your notes to help you conduct a brainstorming session with your partner. After the brainstorming session, write an outline of your investigation. The outline should be in a format similar to that shown in Figure 9.2.

▶ **EXAMPLE OF AN OUTLINE OF STUDENT RESEARCH ON THE COMPOUND.**
FIGURE **9.2**

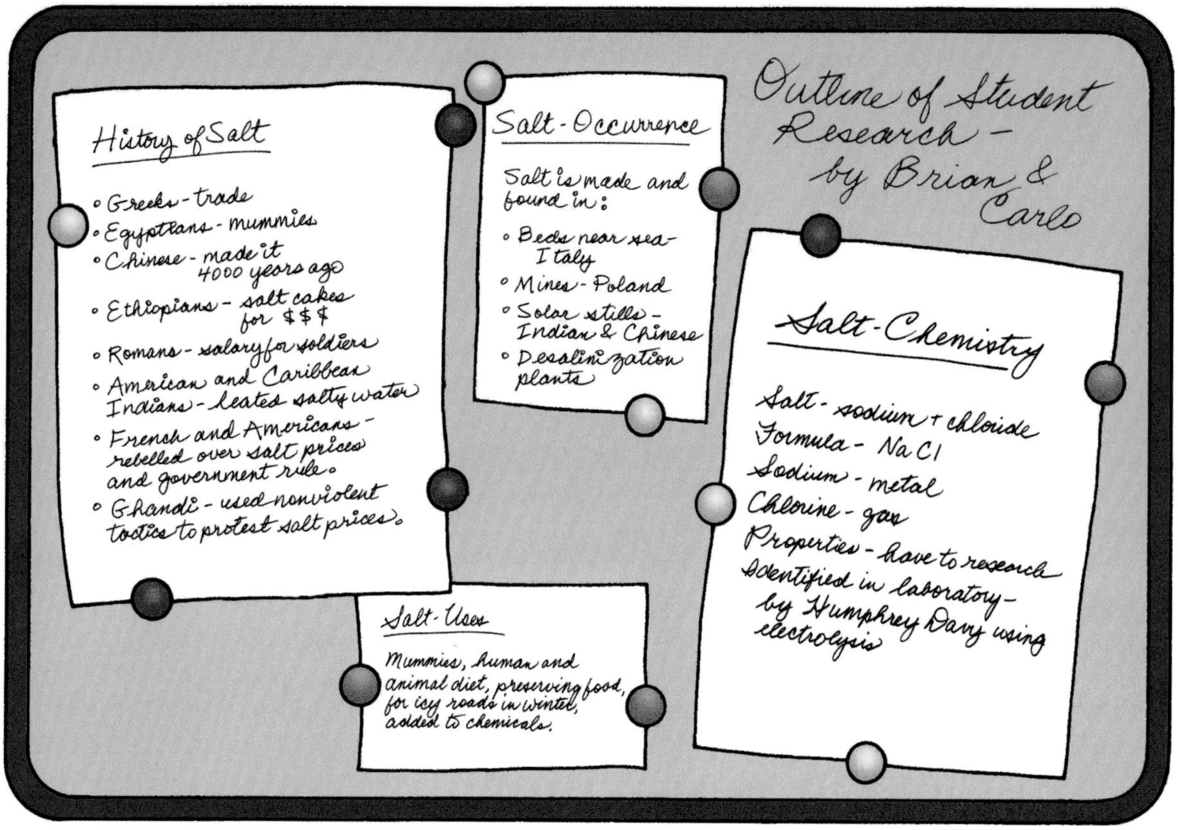

History of Salt

- Greeks - trade
- Egyptians - mummies
- Chinese - made it 4000 years ago
- Ethiopians - salt cakes for $$$
- Romans - salary for soldiers
- American and Caribbean Indians - heated salty water
- French and Americans - rebelled over salt prices and government rule.
- Ghandi - used nonviolent tactics to protest salt prices.

Salt - Occurrence

Salt is made and found in:

- Beds near sea - Italy
- Mines - Poland
- Solar stills - Indian & Chinese
- Desalinization plants

Salt - Uses

Mummies, human and animal diet, preserving food, for icy roads in winter, added to chemicals.

Outline of Student Research - by Brian & Carlo

Salt - Chemistry

Salt - sodium + chloride
Formula - Na Cl
Sodium - metal
Chlorine - gas
Properties - have to research
Identified in laboratory - by Humphrey Davy using electrolysis

Exploration Activity Part 2 *continued*

3 On another sheet of paper, write a bibliography (see Figure 9.3). The bibliography can include books, newspapers, magazines, CDs, DVDs, and TV programs. You should have at least one reputable Internet reference (check with your teacher and/or school librarian about your Internet sources).

4 Hand in your outline and bibliography on or before the due date on your schedule. Your teacher will use this information to make sure your research is heading in the right direction.

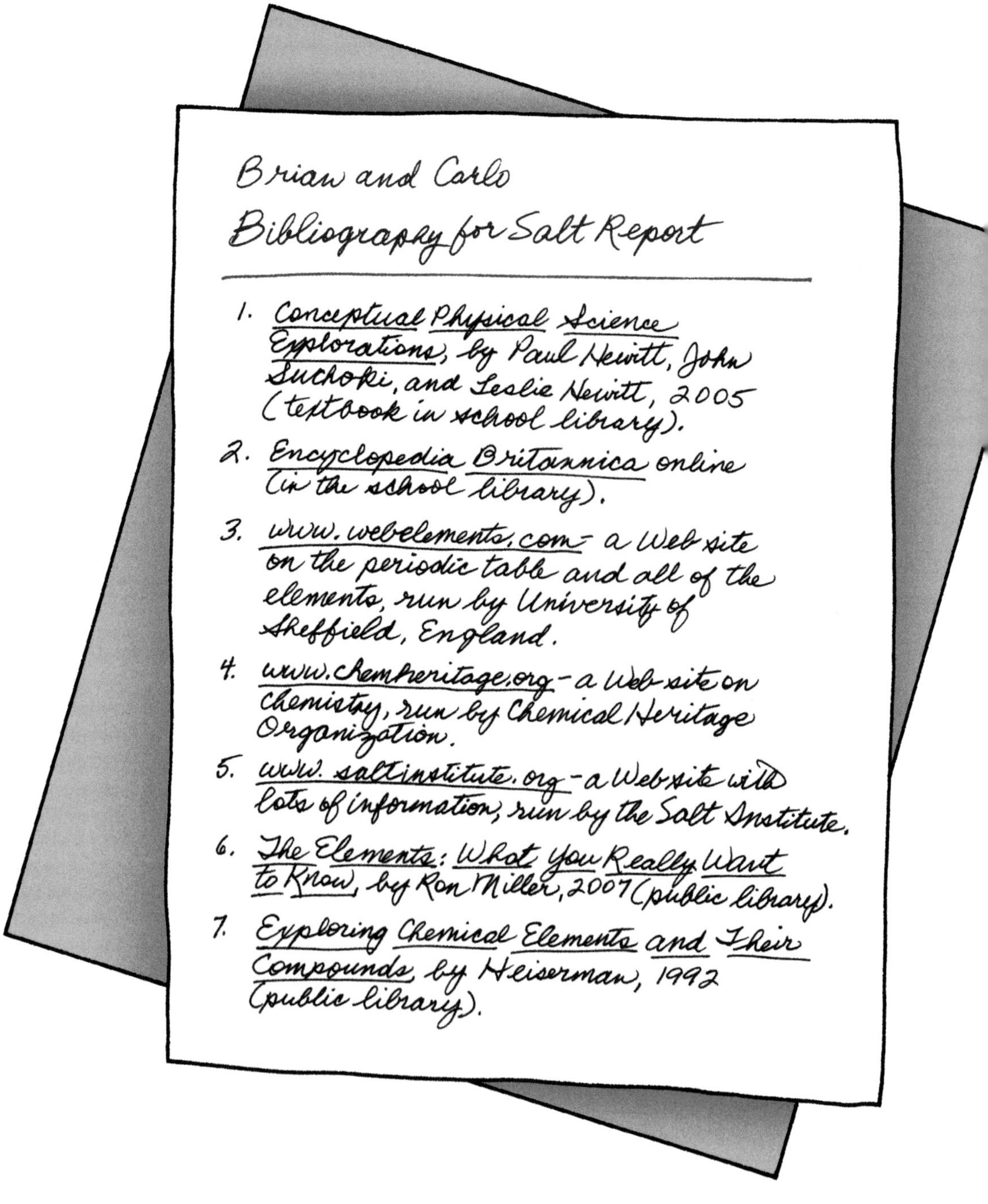

▶ EXAMPLE OF A
BIBLIOGRAPHY OF
STUDENT RESEARCH.
FIGURE **9.3**

Brian and Carlo
Bibliography for Salt Report

1. *Conceptual Physical Science Explorations*, by Paul Hewitt, John Suchoki, and Leslie Hewitt, 2005 (textbook in school library).

2. *Encyclopedia Britannica* online (in the school library).

3. www.webelements.com – a Web site on the periodic table and all of the elements, run by University of Sheffield, England.

4. www.chemheritage.org – a Web site on chemistry, run by Chemical Heritage Organization.

5. www.saltinstitute.org – a Web site with lots of information, run by the Salt Institute.

6. *The Elements: What You Really Want to Know*, by Ron Miller, 2007 (public library).

7. *Exploring Chemical Elements and Their Compounds*, by Heiserman, 1992 (public library).

PART 3

CREATING THE CUBE

PROCEDURE

1. Continue your research. At least a week before the due date for the completed exhibit, gather together all of your information to bring to class when it's assigned.

2. Inquiry Master 9a: Scoring Rubric for the Cube explains what is required for each section of your exhibit. Read it carefully because it tells you how you can earn high scores for your exhibit. Notice that points are awarded for the bibliography and for the presentation and effective design of your exhibit. Remember, points will be deducted for late work.

3. Write each section of your exhibit (use the headings listed in Part 2, Step 1 of the five Exploration Activity Guidelines). If you can, use a word processor to type the final text. If you don't have access to a word processor, have the member of your group with the best handwriting do the writing. You have limited space for each of the sections. Choose the content, including pictures and diagrams, very carefully.

Exploration Activity Part 3 *continued*

4 Construct your exhibit. In making a cube, follow the instructions in Steps 5 and 6.

5 Make the cube from lightweight cardboard, card stock, or poster board. The dimensions of the cube should be about $15 \times 15 \times 15$ cm. Figures 9.4 and 9.5 show how to assemble the cube.

6 Use one side of the cube for each heading (this will use five sides). Make sure that at least four sides of the cube read in the same direction (see Figure 9.6). Put a representation or photograph of your compound (or, if possible, the compound itself) on the sixth side. It is important that your names and bibliography appear somewhere on the exhibit. Figures 9.7 and 9.8 show completed cubes.

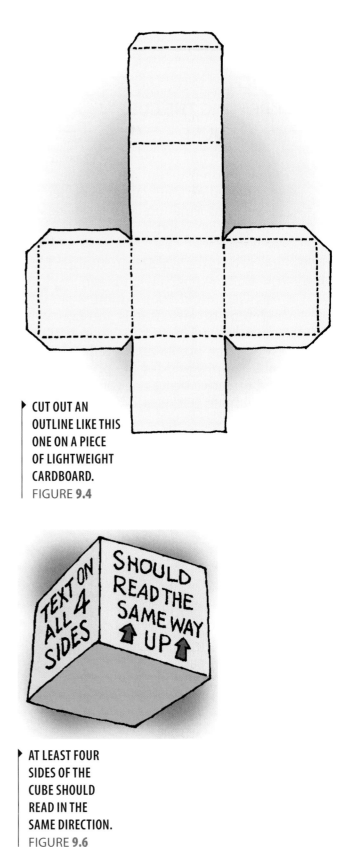

▶ CUT OUT AN OUTLINE LIKE THIS ONE ON A PIECE OF LIGHTWEIGHT CARDBOARD.
FIGURE **9.4**

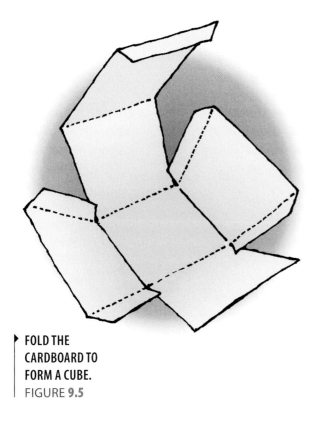

▶ FOLD THE CARDBOARD TO FORM A CUBE.
FIGURE **9.5**

▶ AT LEAST FOUR SIDES OF THE CUBE SHOULD READ IN THE SAME DIRECTION.
FIGURE **9.6**

7 If your cube is too small for all your information to fit, include the additional information in your oral presentation.

8 Hand in your exhibit on or before the due date. Your cube will be exhibited for your classmates and school community to view.

▶ THE STUDENTS WHO BUILT THESE CUBES HAD FUN RESEARCHING AND DESIGNING THEM.

PHOTO: © David Marsland

FIGURE **9.7**

▶ ONCE COMPLETE, THE CUBES FROM YOUR CLASS WILL MAKE AN EXCITING EXHIBIT ABOUT COMPOUNDS AND HOW WE USE THEM. ALTHOUGH EACH CUBE FOLLOWS THE SAME FORMAT, YOUR PERSONAL TOUCHES WILL MAKE YOUR CUBE UNIQUE.

PHOTO: © David Marsland

FIGURE **9.8**

PART 4

GIVING THE ORAL PRESENTATION

PROCEDURE

1. Work with your partner to prepare a short oral presentation. It should focus on the properties, history, occurrence, and uses of the compound you have chosen. You should include the following:

   A. the physical and chemical properties of the compound

   B. the properties of the elements which make up the compound

   C. the place where the compound is found in nature or its creation in the laboratory

2. Both you and your partner should be involved in giving the presentation. During your presentation, use some visual aids, such as computer presentations, posters, maps, or overhead transparencies. You may also use Web pages or a short video.

3. Carefully read Inquiry Master 9b: Scoring Rubric for the Oral Presentation. It tells you how your oral presentation will be assessed. Use the table to plan your presentation.

4 With your partner, practice giving the presentation. Time yourselves so that the presentation is 5 minutes long.

5 Make sure you have all of your materials ready before you give your presentation. You may refer to notes during your presentation, but you should avoid reading them.

WHEN WILL YOU DO ALL OF THIS WORK?

You will be given several homework assignments and some class time to do this work. However, you will have to do most of it on your own time. At the end of the unit, two to three class periods will be used for the Exploration Activity presentations.

A Brief History OF SALT

One of the most valuable substances known to early mankind was salt. It was used as an important part of the human and animal diet, for preserving food, as currency, and in religious rituals and legal ceremonies. Salt played an important role in the political and economic history of many countries. Arguments over the price of salt led to wars and revolutions.

Prehistoric man found salt in ancient lake bottoms, domes, underground deposits, oceans, lakes, and shallow beds near streams. The Egyptians found that deposits of salt along the Nile River were effective in preserving human remains. They used it for the mummification of humans and animals. Even today, large numbers of ancient mummified crocodiles, birds, and cats continue to be uncovered in Egypt.

▶ THE BONNEVILLE SALT FLATS IN NORTHWESTERN UTAH ARE A REMNANT OF AN ANCIENT LAKE THAT EXISTED THERE DURING THE LAST ICE AGE. TODAY, THEY ARE MOST WELL KNOWN FOR THEIR USE AS A HIGH-SPEED RACEWAY. SEVERAL WORLD LAND-SPEED RECORDS HAVE BEEN BROKEN ON THE FLATS.

PHOTO: John Philip Green/
creativecommons.org

▶ THESE MOUNDS ARE ACTUALLY PILES OF SALT ON THE SALAR DE UYUNI SALT FLAT IN BOLIVIA. THE SALAR DE UYUNI IS THE LARGEST SALT FLAT IN THE WORLD, COVERING APPROXIMATELY 8,000 SQUARE KILOMETERS.

PHOTO: Alicia Nijdam/creativecommons.org

Sometimes bodies and objects were accidentally preserved by salt. The bodies of six ancient salt miners were found in an unused part of a salt mine in Iran. These bodies, found from 1993–2006, were referred to as "salt mummies." The salt in the mine preserved the bodies of the miners, probably killed by rockslides in the mines over 2,500 years ago. Their bodies, clothes, tools, and ancient leather bags used to collect the salt were perfectly preserved. Scientists continue to study the bodies for the information they can give us about the past.

Both the Romans and the Chinese got their salt from brinish (salty) water. Records from China show that over 4,000 years ago the Chinese heated seawater in an early version of a solar still. The Romans used solar heat to evaporate water from salty water deposits along the Mediterranean Sea. When their salt beds were covered by the rising Mediterranean Sea, the Romans shipped salt from more distant parts of their empire—present-day England and Turkey—back to Rome.

READING SELECTION

EXTENDING YOUR KNOWLEDGE

In other parts of Europe, salt was dug from mines. One of the oldest and largest is Wieliczka, near Krakow, Poland. Today the mine is a museum, and tourists can walk along some of its 200 km (120 miles) of passages. During the 700 years of its use, the miners made elaborate carvings of political and religious figures. They even sculpted whole rooms into chapels and meeting areas where weddings, funerals, religious services, and important meetings are held today.

The first Europeans traveling to the New World found the Indians in the Caribbean harvesting salt. In North America, salt production was also common among the Indian tribes. The deep vessels used for boiling salty water have been found on Avery Island in Louisiana (although Avery Island is best known for its production of hot pepper sauce rather than salty sauce).

All of the civilizations and groups that produced salt used it for trade and commerce. The Greeks had no local salt production, so they traded their captives of war for salt. The Ethiopians made salt cakes and used them for money to purchase goods from others. The Romans paid their soldiers with a salt allowance, and gave us the term "salary" (from the Latin "sal" for salt) to mean a regular payment of wages.

The need for salt influenced political events in North America, Europe, and Asia. In America, during the colonial period, the British raised the tax on salt, along with taxes on tea and other products used by the colonists. The Americans resented the tax increases on such important items so much that they dumped tea in the Boston harbor to avoid paying the higher tax on it. Eventually those feelings of resentment and mistreatment led to the declaration of American independence and the American Revolution against the British.

At the same time, in France, the government also controlled the sale of salt. It could be bought only in royal stores or depots owned by the nobles who were chosen by the king. The local people resented the government's control over a precious substance and resorted to smuggling and sales on the black market. In 1789, resentment over the French royal government peaked and the people rebelled. The whole depot system ended when the government was overthrown, but some years later Napoleon reinstated the salt tax to raise money for his army. The tax remained in effect until Napoleon was defeated and exiled.

▶ ST. KINGA'S CHAPEL WAS SCULPTED UNDERGROUND
IN THE WIELICZKA SALT MINE.

PHOTO: jhadow/creativecommons.org

READING SELECTION
EXTENDING YOUR KNOWLEDGE

In the 20th century in Asia, a young Indian lawyer led a nonviolent rebellion against the British over their taxes on salt sold to their colonial subjects in India. Although he failed to get tax relief for his countrymen, Mahatma Gandhi started a political movement that eventually led to Indian independence. His political strategy also became a model for future leaders, including Martin Luther King, Jr., who sought rights, freedoms, and civil liberties without direct violence.

Although salt played a major role in individual lives and world history, it wasn't until the 19th century that chemists were able to recognize that salt was a compound and identify its elements. The Englishman Humphry Davy was responsible for the investigation and decomposition of salt in the laboratory. He used a battery, recently invented by an Italian scientist, Alessandro Volta, to run an electric current through molten sodium hydroxide and separate the metal sodium from the hot liquid. (This was similar to the electrolysis experiment that you performed to separate water into hydrogen and oxygen.) Soon after identifying sodium, Davy did other experiments to isolate chlorine gas. Davy was an avid experimenter and separated many compounds into their elements. He was credited with the discovery of barium, strontium, magnesium, calcium, and potassium. He also invented a safe source of light for use by coal miners and was knighted by the British government, and given financial rewards along with his title.

▸ **HUMPHRY DAVY USED ELECTROLYSIS TO IDENTIFY SODIUM AS ONE OF THE ELEMENTS IN SALT.**

PHOTO: Courtesy of Smithsonian Institution Libraries, Dibner Library of the History of Science and Technology, Washington, D.C.

Today, we recognize that salt is a compound of sodium and chlorine. We give it the chemical formula NaCl. The Na represents the very active metallic element sodium; the Cl is the chemical abbreviation for chlorine, which is a poisonous gas in its pure form. When alchemists and chemists first started identifying the elements, they used symbols and Latin terms. Na is the abbreviation for the Latin "natrium," or "soda" in English. An examination of sodium and chlorine shows that the elements have very different properties from the compound that they form when combined. Sodium and chlorine are important elements by themselves and form many other compounds when combined with other elements.

Salt continues to be a highly useful compound in modern times. It plays an important role in human and animal nutrition, for highway safety and mobility, and most importantly as a necessary component in chemical production. Salt is produced commercially on all continents except Antarctica; the United States, China, Germany, Canada, Australia, and Mexico are its largest producers. ■

DISCUSSION QUESTIONS

1. Use library and Internet resources to research why salt is an important component of human and animal nutrition.

2. Salt has long been used in the preservation of meats. Research why salt is used for this purpose and how it works.

CHEMICAL REACTIONS INVOLVING METALS

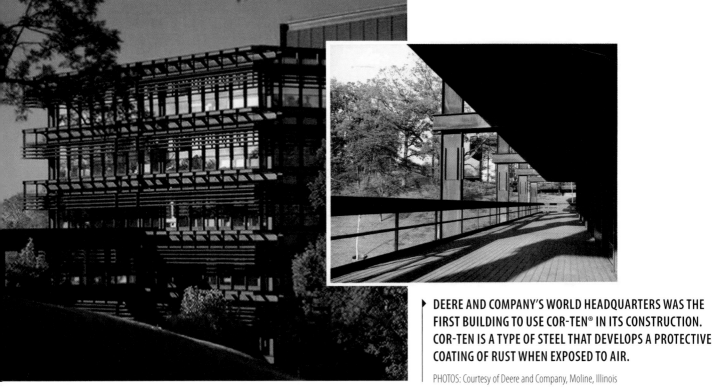

▶ DEERE AND COMPANY'S WORLD HEADQUARTERS WAS THE FIRST BUILDING TO USE COR-TEN® IN ITS CONSTRUCTION. COR-TEN IS A TYPE OF STEEL THAT DEVELOPS A PROTECTIVE COATING OF RUST WHEN EXPOSED TO AIR.

PHOTOS: Courtesy of Deere and Company, Moline, Illinois

INTRODUCTION

The earth's crust contains large amounts of many different metals, but pure chunks of metal are rarely found. Aluminum, for example, is the most common metal in earth's crust, but it is never found naturally as a piece of aluminum metal. Aluminum exists naturally only combined with other elements as aluminum compounds. You know that iron reacts with oxygen in the air. Would you expect to find a lot of pure iron metal lying around? Some metals, such as copper and gold, can be found as nuggets of pure metal. Why is it possible to find a rare metal, such as gold, in the form of pure nuggets, but not a common metal, such as magnesium? In this lesson, you will investigate the differences in the chemical properties of some metals. After this lesson, you may be able to answer some of these questions.

OBJECTIVES FOR THIS LESSON

Conduct an inquiry to compare how different metals react with acid.

Discuss how differences in the chemical properties of metals affect how they are extracted from their ores and used.

Design and conduct an experiment to compare how different metals corrode.

Develop an understanding of the physical and chemical properties of acids and bases from the inquiries and the reading selection on this topic.

▶ MATERIALS FOR LESSON 10

For you

    Your copy of Student Sheet 7.1a: Examining and Grouping Elements

  1   copy of Student Sheet 10.1: Comparing the Reactions of Different Metals with Acid

  1   copy of Student Sheet 10.2: Investigating Corrosion

  1   pair of safety goggles

For you and your lab partner

  1   test tube rack

  4   test tubes

  1   lab scoop

  1   bottle containing dilute hydrochloric acid

  1   thermometer

  1   metric ruler

For your group

  1   jar of magnesium ribbon pieces

  1   jar of granular zinc

  1   jar of copper filings

  1   jar of iron filings

  1   plastic container

  1   black marker

  2   cotton balls

  1   bottle of vegetable oil

  5   labels

     Access to boiled water

     Access to anhydrous calcium chloride

GETTING STARTED

1 You know that metals have many properties in common. But do metals also have properties that differ from one another? Discuss with your group how metals may differ in their properties. Be prepared to contribute your ideas to a class discussion.

2 Use the information you collected in Table 1 on Student Sheet 7.1a and your observations from Inquiry 8.2 to find out how one of the following metals reacts with oxygen in the air: copper, iron, magnesium, sodium, aluminum, zinc, calcium, or tin. Your teacher will assign a metal to your group.

3 Report on your findings. After your teacher lists your results, write the list in your science notebook. You will use this list later in the lesson.

INQUIRY 10.1

COMPARING THE REACTIONS OF DIFFERENT METALS WITH ACID

PROCEDURE

1 One student from your group should obtain a plastic box containing the materials.

2 Check the items against the materials list.

3 Remove the materials that you and your lab partner will use from the plastic box. You will share the jars of elements with the other pair in your group.

4 Pass around the jars and examine the elements. Your teacher will ask you some questions about the characteristic properties of the substances in the jars.

SAFETY TIPS

You must wear your safety goggles *throughout the lesson.*

Immediately inform your teacher of any spills or accidents you have involving acid.

You will be handling dilute hydrochloric acid. If you spill it on your clothes or your skin, wash it off immediately with lots of water. If you get some in your eyes, wash it out with lots of water. If you spill it on the bench, immediately inform your teacher.

5 You will be comparing some chemical properties of these elements. Listen carefully while your teacher reviews Steps 6 through 11 below. Reread the procedure before starting the experiment.

6 Place the four test tubes in the test tube rack.

7 Pour hydrochloric acid into each tube to a depth of 5 cm. Use your ruler to measure the acid in the test tube.

8 Add two lab scoops of iron filings to the first test tube. Immediately place the thermometer in the tube and measure the temperature (see Figure 10.1).

9 Carefully observe what happens in the test tube. After a few minutes, measure the temperature again.

10 Record your observations and temperature measurements in Table 1 on Student Sheet 10.1: Comparing the Reactions of Different Metals with Acid.

11 Repeat Steps 8 through 10 with another metal and a new test tube containing acid. Perform the same procedure with the remaining two metals.

▶ **WATCH WHAT HAPPENS IN THE TEST TUBE. CAREFULLY MEASURE ANY TEMPERATURE CHANGES.**
FIGURE **10.1**

Inquiry 10.1 continued

12 Use your observations and measurements to help you answer the following questions on Student Sheet 10.1:

A. Did any chemical reactions take place?

B. What evidence was there that chemical reactions were occurring?

C. Did all the metals react with the acid?

D. Of those that did react, did they all behave in the same way?

E. Did the products of the reactions look the same?

F. Were there any temperature changes? If so, did the temperature increase or decrease?

G. Were the temperature changes the same for all of the metals?

13 Your teacher will conduct a demonstration with the metals you used in this inquiry.

14 The metal that reacts the fastest with the acid is the most reactive. Use your results from Table 1, your answers to the questions on the student sheet, and your observations of the demonstration to complete Table 2 on Student Sheet 10.1.

15 Is there any similarity between the information in Table 2 and the information you collected in "Getting Started"? Record your answer.

16 Clean up your work area. Pour the acid and any remaining metal into the bucket provided by your teacher. Do not pour it down the sink. Wash out the test tubes with lots of water before returning them, with the other materials, to the plastic box.

INVESTIGATING CORROSION

PROCEDURE

1 In this inquiry you will work in your group to investigate how four metals corrode. Discuss with your group what you think the word "corrode" means. Your teacher will ask you for your definition.

2 Think about the following questions, and then discuss your ideas with your group:

A. Do metal objects that are placed outdoors corrode?

B. What causes corrosion?

C. Do all metals corrode?

3 In this inquiry, you will consider how to design an experiment to compare how typical environmental conditions (that is, exposure to air and water) cause different metals to corrode. The metals you will investigate are the same ones you used in Inquiry 10.1: magnesium, zinc, iron, and copper. Discuss with your group how you could set up an inquiry to determine the effects of air and water on one of these metals. Look at the apparatus in the plastic box to give you some ideas. Think about the following questions:

A. What conditions will you need to create to show that both air and water affect the corrosion of metals?

B. How will you create these conditions?

C. How will you ensure that all comparisons you make are fair?

4 After 5 minutes of group discussion, your teacher will conduct a brainstorming session on your ideas.

Inquiry 10.2 continued

5 Use the class procedure to complete the first two columns of Table 1 on Student Sheet 10.2: Investigating Corrosion.

6 Your group will set up the apparatus to investigate the corrosion of one of the metals. Your teacher will assign a metal to your group. Record the name of the metal.

7 Using the labels and a marker, label each test tube with the numbers used in column 1 of Table 1. Also label the plastic container with your group members' names and the metal you are investigating.

8 As you set up each test tube, stand it in the plastic container. Place the container in the designated storage place (see Figure 10.2). You will look at the results of this experiment in about 4 days.

9 After about 4 days, record your observations in Table 1 on Student Sheet 10.2.

10 Discuss the results with your group and write your conclusions and any other notes you may have for each tube in Table 1.

▸ PLACE THE NUMBERED TUBES IN THE LABELED CONTAINER AND PUT THEM IN THE PLACE DESIGNATED BY YOUR TEACHER.
FIGURE **10.2**

11 Your teacher will help you collect the results for all the metals from different groups. Summarize the class results in Table 2 on Student Sheet 10.2. Write your own conclusions for each metal.

12 Use the class results to answer the following questions on Student Sheet 10.2:

- Did all the metals corrode?

- Did all the metals corrode to the same extent?

- What is the relationship between the rate at which a metal corrodes in the presence of air and water and the rate at which it reacts with acid?

REFLECTING
ON WHAT
YOU'VE DONE

1 Review the results from both inquiries as well as the information you collected in "Getting Started." Do you recognize any similarities between the data obtained from the two inquiries? Be prepared to contribute your ideas about the chemical reactivity of metals to a class discussion.

2 On Student Sheet 10.2, write a short paragraph about what you have found out about the chemical reactivity of metals (see Figure 10.3) and how this knowledge can be applied to choosing metals to do specific jobs (for example, the use of copper to make water pipes).

3 Read "About Acids and Bases" on page 128 and answer the questions. Compare your answers with those of your classmates.

▶ **WHICH METAL IS THE MOST REACTIVE?**
FIGURE **10.3**

About ACIDS AND BASES

Acids and bases were known and used by alchemists before their chemical structures were identified. Substances such as vinegar and citrus fruit juice were identified as sour tasting and described as acids. The term "acid" is taken from the Latin "acidus" for "sour." French chemist Antoine Lavoisier identified the gas produced by the reaction of acids with metals. Other substances known for their bitter taste and slippery feel were called "alkalis." The word "alkali" came from the Arabic word for "ashes" because these substances were obtained by soaking ashes from wood fires in water.

In the late 1800s, Swedish chemist Svante Arrhenius was testing the ability of certain solutions to conduct electricity. (The invention of a storage battery to produce electricity led to many electrolysis experiments.) He theorized that some solutions, called electrolytes, decomposed into charged particles, or ions, and produced electricity when dissolved in water. Acids, he suggested, split up when placed in water and produced hydrogen ions with a single positive charge (H^+). He described bases as substances that dissolved in water and formed negatively charged hydroxide ions (OH^-).

Some common acids and their chemical formulas are hydrochloric (HCl), nitric (HNO_3), sulfuric (H_3PO_4), carbonic (H_2CO_3), and acetic ($C_2H_4O_2$). All carbonated beverages contain carbonic acid, and many have small quantities of phosphoric acids. Citrus fruits include ascorbic acid, or vitamin C ($C_6H_8O_8$). Many household cleaning products contain hydrochloric acid. You will notice that all these compounds contain hydrogen, which produces the Arrhenius H^+ ion in water.

Bases are equally common in everyday life and use. They include sodium hydroxide ($NaOH$) (often called lye and sometimes used as a drain cleaner), ammonia (NH_3), baking soda ($NaHCO_3$), substances contained in ashes (KCO_3), and soaps. When solutions of bases and acids are mixed, water and a new substance are produced. This reaction is called a neutralization reaction because the acid (H^+) and the alkali (OH^-) neutralize one another to form H_2O. The new substances formed in neutralization reactions are called "salts." One of these salts, $NaCl$, the product of the reaction between hydrochloric acid and sodium hydroxide, is what we call "table salt," or sodium chloride. This is only one of many substances called salts.

Arrhenius's definitions of acids and bases were modified in the 20th century as the understanding of atomic structure grew. Two scientists working independently reached the same conclusion on the structure of acids and bases. Chemists T.M. Lowry from England and J.N. Brønsted of Denmark each developed new definitions of acids and bases. They defined an acid as a substance that can donate a proton, the H^+ ion. They defined a base as a substance that can accept a proton. This definition of acid-base reactions, known as the Brønsted-Lowry Theory of acids and bases, is widely accepted by scientists today. The Brønsted-Lowry Theory illustrates the gradual progression of explanations in science. New understandings of the structure of matter have brought new explanations for well-known behaviors of matter. ∎

DISCUSSION QUESTIONS

1. How do acids and bases differ in their chemical and physical properties?

2. Why do scientists change their explanations of things over time?

Reactivity and Free Metals

Gold is valued for its scarcity and its lack of reactivity. Because gold does not react with other elements, such as oxygen, it can be found in its pure form as nuggets. Metals that are not combined with other substances are called free metals. Gold's lack of reactivity also means that it stays shiny and does not corrode. This makes it ideal for use in jewelry and in electrical contacts (for example, the plugs on a computer cable).

▶ **THE LOW REACTIVITY OF GOLD EXPLAINS WHY IT CAN BE FOUND AS METAL NUGGETS. MOST OTHER METALS EXIST AS ORES.**

PHOTO: U.S. Geological Survey

READING SELECTION

EXTENDING YOUR KNOWLEDGE

▶ COPPER IS MORE REACTIVE THAN GOLD. SOMETIMES IT IS FOUND AS NUGGETS AS SHOWN IN THIS PICTURE, BUT MOST COPPER IS OBTAINED BY SMELTING COPPER ORES. DURING SMELTING, THE METAL IS EXTRACTED FROM THE METAL ORE. THE EXACT PROCESS DEPENDS ON THE TYPE OF ORE. FOR EXAMPLE, IF THE ORE IS COPPER CARBONATE OR COPPER OXIDE, SMELTING IS ACHIEVED BY ROASTING THE ORE WITH CARBON, USUALLY IN THE FORM OF COKE. A CHEMICAL REACTION TAKES PLACE THAT PRODUCES COPPER METAL AND CARBON DIOXIDE GAS.

PHOTO: Ken Hammond/USDA

▶ THIS IRON METEORITE ON MARS IS IN THE FORM OF PURE IRON, WHICH IS RARELY FOUND ON EARTH. WHY CAN PURE IRON EXIST IN SPACE BUT NOT FOR VERY LONG ON THE SURFACE OF EARTH?

PHOTO: NASA/JPL/Cornell

Copper is a much more common metal than gold, and sometimes nuggets of pure copper can be found. Copper can remain shiny for a long time, and it is also used to make jewelry. However, eventually it reacts with other elements, particularly oxygen in the air, and slowly corrodes (a process called tarnishing).

Iron is more reactive than copper. Because of its reactivity, iron is almost always found in the earth's crust combined with other elements. Rocks containing large amounts of iron compounds are called iron ore. Iron is extracted from these ores by a process called smelting. Chunks of natural iron are sometimes found, but these originate from outer space. Meteorites that fall to earth are often composed mainly of iron. They can last thousands of years, but they also eventually corrode. ■

? DISCUSSION QUESTIONS

1. Although iron is one of the most abundant metals in the universe, and we might expect Earth to contain a great deal of it, why is it rare to find big chunks of iron?

2. Research another metal (besides gold, copper, or iron) and describe its properties relative to extraction and corrosion.

PANNING *for Gold*

> **THE PROSPECTOR SHOWN IN THE PHOTOGRAPH ABOVE IS TRYING HIS LUCK AT PANNING FOR GOLD.**

PHOTO: Library of Congress, Prints & Photographs Division, LC-USZ62-120295

Gold is a very popular metal. It has a beautiful yellow luster. It is soft and easy to shape into jewelry. It conducts electricity very well, so it is great for use in electronic devices. It doesn't rust or tarnish, so it is always shiny. It is also very rare. Fortunately, for those who want to find gold, it is extremely dense. Gold prospectors use the great density of gold to separate rare gold flecks from ordinary rocks, pebbles, sand, and silt.

HARD WORK

Panning is the simplest technique for finding gold in streams and rivers. All that is needed is a simple flat pan, a strong back, and patience. The prospector shown in the photograph above tried his luck in the Yukon gold rush.

SPEEDING THINGS UP

Gold rush prospectors (like the ones shown in the drawing above) found gold in rivers and streams, because rain had washed it there from surrounding hills over thousands of years.

The dense flakes of gold settled quickly in streams and were caught in nooks and crannies on the bottom. Panning worked well at first. But once all of the easy pickings were gone, prospectors began to use water jets to wash whole hillsides away. They then used long sluices to search through the washings. Sluices are wooden troughs with shallow grooves called riffles in the bottom. The gold would get stuck in the riffles as water and mud washed down the sluice. Every now and then, the prospectors would stop the flow and remove the gold. Because this method of prospecting caused large amounts of silt to pollute rivers, it was restricted in California as early as the 1880s.

DENSITY MAKES IT POSSIBLE

Only a small portion of gold washed into a river is in the form of nuggets big enough to see easily. Most gold in a stream is dust—tiny flecks mixed in with ordinary dirt. These dust particles aren't heavy; each one weighs only a fraction of a gram. But gold particles, no matter how small, are the densest part of the mix. This causes the gold flecks to sink through water faster than everything else.

▶ GOLD NUGGETS FOUND IN 1910 IN AN ALASKAN CREEK.

PHOTO: U.S. Geological Survey

▶ GOLD MINING IS STILL COMMON IN MANY REGIONS OF THE WORLD. THIS
PROSPECTOR IS PANNING FOR GOLD IN AFRICA.

PHOTO: Michael Modzelewski/Adventures Unlimited

To prospect for gold, prospectors use a pan to dredge up silt and rocks from the bottom of a river. They pick out the big rocks and add water from the river to the pan (see the photograph above). Then they swirl the pan of silt and water, allowing anything that does not sink quickly to the bottom of the pan to flow out of the pan.

Prospectors repeat this process until they are left with only "black sand," which is very dense. If they are lucky, the black sand will contain tiny gold flecks. Then they pick out the flecks, and start again. If they are very lucky, they will find nuggets of gold (see the photograph at left). ■

DISCUSSION QUESTIONS

1. What is gold used for today? Why is gold an appropriate metal for those uses?

2. Use library and Internet resources to investigate other methods of gold extraction. Also determine which countries produce gold and gold's main uses.

Making Metals by Mistake

Are you the sort of person who does things by trial and error? For example, if you get a new video game, do you put it in your machine and try to play without reading the instruction manual? If so, you are in good company. Trial and error is a technique that has been used since prehistoric times. It is still used today because it works.

Consider the way people figured out how to make metals. Before about 6500 B.C., the only metals that people used came from nuggets of gold, silver, and copper. Nobody knew that metals were locked inside rocks. Nobody, that is, until someone probably made a very hot fire on top of some rocks and saw molten metals, such as lead or tin, trickling out of the rocks, or ores. These are metals that can be produced at the relatively low temperatures of campfires.

Early people later found that they could use very hot fires to extract copper from its ores or from minerals found in rocks. This discovery probably was accidental—possibly when early potters decorated their pots with minerals containing copper. When the pots were fired in a very hot kiln, the copper melted out of the ore and formed on the pot.

Today, it is known that those ores contained copper, combined with oxygen and other elements. Heating the ores with a very hot fire to several hundred degrees caused a chemical reaction. The ore turned to copper oxide, and the oxygen broke loose from the copper and combined with carbon from the wood. Carbon dioxide floated away. The copper stayed behind.

Of course, early people didn't know any of this. But that didn't stop them from experimenting. How much rock should be used? Which rocks work? Where are the rocks found? How hot does the fire have to be? Does the type of wood matter? Does the phase of the moon make a difference?

▶ EARLY TOOLS, SUCH AS THIS AXE, WERE MADE FROM ALLOYS OF LESS REACTIVE METALS. BRONZE, AN ALLOY OF TIN AND COPPER, WAS USED TO MAKE THIS AXE SOMETIME DURING THE SHANG DYNASTY OF CHINA.

PHOTO: Freer Gallery of Art, Smithsonian Institution, Washington, D.C.: Purchase, F1946.5

BRONZE IS HARDER THAN EITHER TIN OR COPPER— HARD ENOUGH TO USE FOR ARMOR IN BATTLE.

Would it help to add some dirt? If rocks could be changed into copper, could copper be turned into gold? After a few centuries of trials—and almost as many errors—people knew a lot about making copper, and it became widely available.

The copper was not very hard. It could be made into pots and pans. It could be shaped into fancy jewelry. But it was too soft to make good tools or weapons. People needed tools and weapons, and more trial and error eventually led to the next big discovery.

Sometime around 3800 B.C., a copper maker in the Middle East mixed tin ore with copper ore and heated them up. The resulting metal was very different from tin and from copper. This new metal, an alloy called bronze, was lighter in color than copper. It was also much harder than either copper or tin. Bronze eventually was used to make axes, spears, knives, armor, and other tools.

The secrets of making bronze also emerged in the Far East. By 1500 B.C., Chinese bronze makers had discovered, by trial and error, that the hardest bronze is exactly 85 percent copper and 15 percent tin. They had no idea why this particular mixture was so hard, but by experimenting, watching carefully, and recording results, they found the best way to make bronze.

Iron was probably discovered by mistake, in much the same way as copper. However, iron ore requires a much hotter fire than that used to extract copper from ore. Can you think of the reason why the fire needs to be so hot?

Iron is much harder than bronze. Tools and weapons made of iron were much harder than those made from bronze. The techniques for extracting and improving the quality of iron were refined through trial and error, and the new technology spread quickly. The Iron Age had begun. ■

DISCUSSION QUESTIONS

1. Identify two objects made from iron. For each object, state why iron, rather than any other material, was used in its manufacture.

2. Iron was discovered after copper and tin because it is more difficult to extract from its ore. What are the modern processes for extracting iron metal from its ores? Use Internet and library resources to learn more about these processes, and draw a diagram to illustrate the techniques involved.

COUNTERING CORROSION

INTRODUCTION

Imagine you possessed the scientific knowledge to save the country tens of billions of dollars every year! If you could prevent a particular process from occurring, that much money could really be saved. The process is a chemical reaction that corrodes objects made of iron and steel. What is the common name given to this process? Can you identify the reactants and the products in this chemical reaction or the conditions needed for it to take place? How could you prevent it? In this lesson, you will investigate this process and find out the answers to these questions.

▶ **WHY IS IT IMPORTANT TO PAINT A STEEL SHIP?**

PHOTO: National Archives and Records Administration

OBJECTIVES FOR THIS LESSON

▸ Discuss ideas about the nature and causes of rusting.

▸ Design and conduct an inquiry to compare the effectiveness of different rust-prevention techniques.

▸ Explain results in terms of the chemical reaction involved in the rusting process.

▸ **MATERIALS FOR LESSON 11**

For your group

1	black marker
2	petri dishes with lids
3	ungalvanized steel nails
1	ungalvanized painted steel nail
1	galvanized steel nail
1	stainless steel nail
1	magnesium ribbon
1	paper towel
2	labels
	Access to a jar of petroleum jelly

GETTING STARTED

 1 Your teacher will show you some objects. Discuss them with the class.

2 Discuss the answers to the following questions with your group:

A. What is rust?

B. What conditions are required for rusting to take place?

C. Why is rusting a problem?

3 Think of as many methods of rust prevention as you can. After a few minutes, your teacher will conduct a brainstorming session on your ideas about preventing rust. Record these ideas in your science notebook. 🖝

▶ **HOW DO YOU THINK THIS BARBED-WIRE FENCE WILL BE AFFECTED BY RUST OVER TIME?**

PHOTO: Ray Elliott/creativecommons.org

INQUIRY

CAN RUSTING BE STOPPED?

PROCEDURE

1 In this inquiry, you will work with your group to design an experiment to compare the effectiveness of different methods of rust prevention.

2 One member of your group should obtain a plastic box containing the materials. Check its contents against the materials list.

3 Write the title of the inquiry in your science notebook. Under the title, write a sentence or a short paragraph describing what you are trying to find out. 🖝

4 Discuss with your group how you could use the materials in the plastic box to design an experiment to compare the effect of corrosion on the following objects: a nail wrapped in magnesium, a stainless steel nail, a nail treated with paint, a galvanized (zinc-coated) nail, and a nail coated with petroleum jelly.

5 Come to an agreement on the design of your experiment and then set up the experiment. Using the labels in your plastic box, label your experiment with the names of the members of your group.

6 Draw a labeled diagram showing how you set up the experiment. Below the diagram, write a short description of the procedure.

7 Design and draw your own results table. You will need to record the appearance of the nails at the start of the experiment and at the end (which will be after at least 3–4 days). You may want to rate each nail's "rustiness."

8 Make sure that you check your apparatus every class period during the course of the experiment. It is your group's responsibility to make sure that the conditions the objects are exposed to remain constant.

REFLECTING
ON WHAT
YOU'VE DONE

1 Discuss the class results with your group. Think about the answers to the following questions:

A. Did all the rust-prevention techniques work?

B. Were they equally effective?

C. Why did some techniques work while others did not?

D. How did the different techniques prevent the rusting reaction from taking place?

2 After the class discussion, write one sentence in your science notebook explaining what happened to each nail.

3 Write a paragraph summarizing everything you know about the process of rusting. Include a word equation for the process.

9 After the designated time, record your results in the table and discuss your findings with your group.

10 A member of your group will be asked to report on some of your findings. Make sure you are ready to make such a report.

The Work Never Ends

The Golden Gate Bridge is one of the world's great bridges. It spans 2.7 kilometers (1.7 miles) across the Golden Gate Strait at the end of San Francisco Bay. More than 1.8 billion cars have crossed over it. It contains 75 million kilograms (about 165 million pounds) of steel. It can withstand powerful tides, high winds, and major earthquakes. But perhaps the greatest threat to the bridge is corrosion. The iron in steel combines with the oxygen in air to form iron oxide, or rust. Water and salt, which the bridge is constantly exposed to, greatly speed up rust formation. To keep the mighty bridge from crumbling apart, maintenance crews must constantly battle rust.

▶ THE GOLDEN GATE BRIDGE (SAN FRANCISCO, CALIFORNIA) IS AN EXAMPLE OF STEEL USED IN CLOSE PROXIMITY TO SALT WATER.

PHOTO: Justin Beck/ creativecommons.org

PREVENTING RUST

Dry steel rusts very, very slowly. But wet steel rusts very quickly. If steel is in contact with salt water, it rusts even faster. Paint prevents rust by keeping oxygen, water, and salt away from the steel surface.

APPLYING THE PAINT

A painter puts down a primer coat. Primer sticks to the metal, but it isn't that tough or waterproof. So after the primer dries, the painter puts on two protective topcoats of paint that keep out air and water.

TOUCHING UP

Whenever rust spots appear on the bridge, maintenance crews go to work scraping off the rust and worn paint. Metal workers then cut out and replace any corroded metal.

REPAINTING

Although it is called "golden," the Golden Gate Bridge has always been painted International Orange. The bridge was first painted in 1937, the year it opened. The original coat of paint lasted for 27 years, requiring only minor touch-ups. Eventually, the metal became so corroded that the entire bridge was repainted. Between 1965 and 1995, bit by bit, all of the old paint was sandblasted off and a new, tougher acrylic formula was put on. But even the new paint can break down in spots. Maintenance crews constantly inspect the bridge and touch up any rusted areas. ∎

DISCUSSION QUESTIONS

1. What are some reasons the Golden Gate Bridge is repainted so frequently?

2. How do car or bicycle manufacturers stop or reduce rusting? Sketch a car or a bike and label all the different methods of rust prevention you can identify.

Recycling:
It's Not Over 'Til It's Over

PHOTO: Wikimedia Commons

Do you recognize this symbol? You may have seen it printed or stamped on many common objects, including cars, computers, plastic or glass bottles, food and drink cans, or cardboard boxes. The symbol points out that the object can be recycled—that is, reused in some way or processed into a new object or a different material.

Why recycle? Two main reasons. First, recycling reduces the need for landfills. There are more than 3,000 active landfills and 10,000 old landfills in the United States. Finding new places for landfills is becoming harder. Landfills must be located near population centers, but few people want to live close to one. Second, recycling reduces the need to mine or harvest natural resources. Recycling paper means cutting down fewer trees. Recycling yard waste means using less fertilizer. Recycling aluminum cans means using less energy to produce aluminum from ore.

Using less energy means a reduction in burning fossil fuels and reducing the carbon dioxide and other greenhouse gases produced. Recycling is definitely one way to reduce human impact on global climate change.

Whether your neighborhood does curbside recycling, delivers to "drop-off" centers, or receives deposit money when bottles and cans are returned, chances are that your community has already begun recycling. Many communities no longer take all of their trash to a landfill. Instead, they separate plastics, aluminum and other metals, paper, glass, and even appliances. They take these materials to recovery facilities to begin the recycling process. Today, the United States recycles more than one-quarter of all its waste. But we still have a long way to go. Some cities recycle about 50 percent of their waste! This gives the rest of us a great goal, but even that record can be better.

▶ SOME COMMUNITIES HAVE BINS WHERE RESIDENTS CAN SORT AND DEPOSIT VARIOUS MATERIALS FOR RECYCLING.

PHOTO: National Science Resources Center

Paper makes up about 35 percent of our waste. Most types of paper waste can be easily recycled. Paper recyclers shred discarded paper and cardboard, add water, mash the mixture into a thick pulp, and pass the pulp through a filter and a press to remove the water. They then heat the paper fibers to finish the process. The fibers are now ready to be processed into usable products.

Eleven percent of our waste is plastic. Plastics are synthetic materials made from petroleum. Unfortunately, we recycle only a small fraction of discarded plastics. Recycled plastics are broken up and melted so they can be made into new objects—like more bottles and jugs! They can even be mixed together to form composite building materials.

▶ BALER MACHINES ARE USED AT RECYCLING CENTERS TO PACKAGE MATERIALS FOR EASY TRANSPORT TO OTHER FACILITIES FOR PROCESSING.

PHOTO: Provided by City Carton Recycling

Recycling steel and aluminum is especially productive. Steel is the most recycled material in the United States. Since magnets attract steel, it is easy to separate steel from other wastes. Steel from cars, old buildings, and appliances can be recycled. So can the steel cans that many of our foods are stored in. These cans are sometimes also called tin cans, but most of the can is made from steel. There is only a thin layer of tin on the inside to protect the food. During the recycling process, this tin is removed. All new steel materials are made using some recycled material.

Although aluminum recycling is more common than ever, we still waste too much aluminum. In fact, between 1990 and 2000 we put enough aluminum in landfills to make more than 300,000 large airplanes! Recycling aluminum saves more than the material from the earth. It takes 95 percent less energy to make a recycled can than it does to make one from scratch. Recycled aluminum is back in the store, as a new can, in about 90 days!

Aluminum and steel recycling use a similar method. The metals are first crushed into large bricks. Then, the bricks are heated in a large furnace until they melt. Once melted, the steel or aluminum is reshaped into sheets called ingots, which can then be used to make new materials.

Some states forbid yard waste—such as grass clippings and leaves—in landfills. Instead, they support another kind of recycling program: composting. This is a special type of recycling that can be done on a small scale in your own backyard or on a large scale at community facilities. The first step in composting is mixing the yard waste together. Naturally occurring insects, bacteria, and worms break down the waste into a material rich in nutrients that can be used as a fertilizer. Composting eliminates a lot of trash that would end up in landfills and reduces the need for chemical fertilizers.

Does your family or school recycle? If not, maybe it's worth finding out how to become part of your local recycling effort. ■

AMOUNT OF WASTE RECYCLED FROM 251 TONS IN 2006

WASTE PRODUCT	PERCENTAGE OF WASTE RECYCLED FROM ORIGINALLY GENERATED PRODUCT (%)
PAPER	51.6
YARD TRIMMINGS	62.0
FOOD SCRAPS	2.2
PLASTICS	6.9
METALS	36.3
RUBBER, LEATHER, AND TEXTILES	13.3
GLASS	25.3
WOOD	9.4

Environmental Protection Agency Report for 2006

DISCUSSION QUESTIONS

1 Use library and Internet resources to research the process used to recycle glass. How does it compare with the process used to recycle plastic?

2. What are some challenges to the success of a community recycling program?

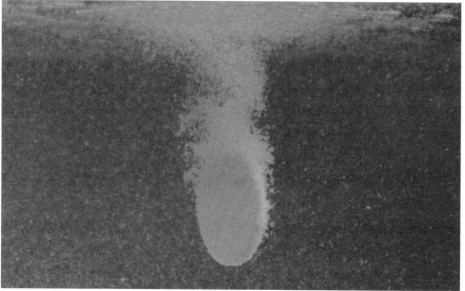

LESSON 12

MASS AND CHEMICAL REACTIONS

▶ IN THIS LESSON, YOU WILL STUDY WHAT HAPPENS TO THE MASS OF REACTANTS AND PRODUCTS IN THIS CHEMICAL REACTION.

PHOTO: National Science Resources Center

INTRODUCTION

In this lesson, you will investigate what happens to the mass of matter when it undergoes change. What rules apply to phase change and to dissolving? Can the same rules be applied to chemical reactions? For example, what do you think happens to the mass of a candle when it is burned? What happens to the mass of a nail when it rusts? What do you think happens to the total mass of matter in a chemical reaction when one of the products is a gas (for example, when you reacted magnesium with acid)? Is the mass of the reactants of these chemical reactions the same as the mass of the products? In Inquiry 12.1, you will investigate mass changes in a chemical reaction. You will measure mass before and after adding half an effervescent tablet to a beaker of water. You will then perform the same experiment inside a sealed bottle. You will be asked to interpret any changes in mass that take place in both experiments and to decide whether the law of conservation of mass can be applied to chemical reactions.

OBJECTIVES FOR THIS LESSON

Conduct an inquiry to compare the mass of the reactants and the mass of the products in the chemical reaction that takes place when an effervescent tablet is added to water in both open and closed containers.

Determine whether the law of conservation of mass can be applied to chemical reactions.

▶ **MATERIALS FOR LESSON 12**

For you

1 copy of Student Sheet 12.1: Measuring Mass in a Chemical Reaction

1 copy of Student Sheet 12: Review for *Experimenting with Mixtures, Compounds, and Elements*

1 pair of safety goggles

For you and your partner

1 250-mL beaker
1 effervescent tablet
1 paper towel
1 clear plastic soda bottle with a screw cap
   Access to an electronic balance
   Access to water
   Access to a wall clock

GETTING STARTED

1 Your teacher will conduct a review of previous lessons. After the review, write in your science notebook your own definition of the law of conservation of mass. 🖉

2 Discuss with your partner whether the law of conservation of mass can be applied to chemical reactions. Think about a chemical reaction you have observed. The following are some questions you should consider during your discussion:

A. What were the reactants and the products of the reaction?

B. Were they all the same phase of matter? Do you think this affects the mass of the matter?

C. You observed a burning candle in Lesson 1. What do you think happened to the mass of matter in that chemical reaction?

3 Your teacher will ask you about some of your ideas. While you are doing Inquiry 12.1, think about the points raised during this discussion and how your ideas change in response to the experimental evidence.

SAFETY TIP
Wear your safety goggles throughout the inquiry.

▶ **WHAT IS HAPPENING TO THE MASS OF THIS CANDLE AS IT BURNS?**

PHOTO: Darren Copley/creativecommons.org

MEASURING THE MASS OF REACTANTS AND PRODUCTS

PROCEDURE

1 One member of your group should obtain a plastic box of materials. In this lesson, you will work with your lab partner. Each plastic box contains the materials for two pairs of students. Make sure your pair has the apparatus indicated in the materials list.

2 Pour about 50 mL of water into the beaker. If you get the outside of the beaker wet, dry it using a paper towel.

3 Break the tablet in half. Shake or blow off any loose grains of tablet.

4 Go to the balance. Make sure the pan of the balance is dry. If not, gently wipe it dry with a paper towel. Check that the balance reads 0.0 g.

5 Place half of the tablet next to the beaker of water on the balance (see Figure 12.1). Record their total mass in column 2 of Table 1 on Student Sheet 12.1: Measuring Mass in a Chemical Reaction.

▶ MAKE SURE THE PAN OF THE BALANCE IS DRY BEFORE FINDING THE MASS OF THE BEAKER, WATER, AND HALF OF THE DRY TABLET.
FIGURE **12.1**

Inquiry 12.1 *continued*

6 Remove the beaker and tablet from the pan. Return to your lab station with the beaker and the tablet. Carefully place the tablet in the beaker without splashing or getting your fingers wet. Observe what happens (think back to Inquiry 1.7). Record your observations in column 5 of Table 1.

7 Wait until you can see no further reaction taking place in the beaker (about 2-3 minutes).

8 Return to the balance and measure the mass of the beaker and its contents. Record the mass in column 3 on Table 1. Then calculate any change in mass and record it in the appropriate place in column 4.

9 Repeat the procedure with the other half of the tablet, but this time use the plastic soda bottle instead of the beaker.

A. Measure the mass of the tablet half, the bottle containing 50 mL of water, and the cap on the balance. Record the mass on Table 1.

B. Add the piece of tablet to the bottle, and immediately screw the cap on very tightly. Watch what happens in the bottle and record your observations on Table 1.

C. When the reaction stops, measure the mass of the sealed bottle (see Figure 12.2) and record your result in Table 1.

D. Calculate any change in mass and record it in the table.

▶ AFTER THE REACTION HAS STOPPED, MEASURE THE MASS OF THE BOTTLE, ITS CAP, AND ITS CONTENTS, KEEPING THE CAP TIGHTLY SEALED.
FIGURE **12.2**

10 Unscrew the bottle cap. What do you hear? Why did it make that noise?

11 Allow the open bottle to stand for 2-3 minutes. Place the bottle, the solution, and the cap on the balance and measure their mass.

12 Record your results at the bottom of column 3 in Table 1. Calculate the difference in mass between the open bottle and the closed bottle. Record your measurement in the table.

13 Rinse the beaker and the bottle. Return them to the plastic box.

14 Examine the data in Table 1. Record your answers to the following questions:

- What evidence do you have that a chemical reaction took place when you added the tablet to the water?

- What happened to the mass when the experiment was performed in an open beaker? Try to explain this result.

- Why didn't you get the same result when you used the closed bottle?

- What happened to the mass of the bottle, cap, and contents after the cap was removed? Explain this result.

REFLECTING
ON WHAT
YOU'VE DONE

1 Using your results and your knowledge of the precision of your apparatus, do you think that the law of conservation of mass applies to chemical reactions? Write your answer in your science notebook.

2 Can you formulate your own law of conservation of mass that applies to phase change, dissolving, and chemical reactions? Discuss possible wording of such a law with your partner. Write your definition on Student Sheet 12.1. Your teacher will ask you to contribute your ideas to a class discussion.

3 At the end of the class discussion, write the class definition of the law of conservation of mass on Student Sheet 12.1.

THE MASS OF MATTER

Matter may change from a solid to a liquid. Elements may react together to form compounds. What happens to the mass of matter in a bowl of water when it is left to stand in the hot sun? What happens to the mass of matter in a piece of paper when it is burned? Sometimes in situations like this it seems as if matter is disappearing. But the disappearance of matter is an illusion.

Matter may change from one form to another. For example, when the water in the bowl absorbs energy from the sun and evaporates, it becomes water vapor in the atmosphere. The piece of paper gives off heat and light energy as it burns, and the matter in it is converted into carbon dioxide, water vapor, and other gases that escape into the atmosphere. Some of the mass will remain behind as ash. In both cases, the matter changes its form, but its total mass stays the same. The same mass of each element is present before and after the change. Matter is neither created nor destroyed during these changes.

It took early scientists hundreds of years of scientific study before the law of conservation of mass became accepted. For a long time, scientists had suspected that matter could not be created or destroyed, but nobody had performed an experiment that proved it.

During the late 18th century, French chemist Antoine Lavoisier and his wife Marie-Anne conducted several experiments that demonstrated the conservation of mass. They were famous for accurate observations and insistence on careful measurements. They used accurate balances that could measure very small changes in mass during their experiments.

Many of the Lavoisiers' experiments were conducted in sealed glass containers from which matter could not escape or enter. For example,

▶ ANTOINE LAVOISIER (1743–1794) WAS ONE OF THE FOUNDERS OF MODERN CHEMISTRY.

PHOTO: Courtesy of Smithsonian Institution Libraries, Dibner Library of the History of Science and Technology, Washington, D.C.

in one experiment, they put fruit into a sealed container, measured its mass, and then left it in a warm place for a few days. The fruit rotted and changed into a putrid mess. Gas was released from the decomposing fruit and droplets of water formed on the glass, but nothing escaped from the container. Lots of changes had taken place, but the mass of the sealed container and the rotten fruit was equal to the mass measured at the beginning of the experiment.

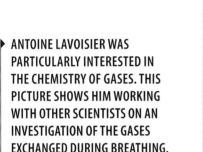

▶ ANTOINE LAVOISIER WAS
PARTICULARLY INTERESTED IN
THE CHEMISTRY OF GASES. THIS
PICTURE SHOWS HIM WORKING
WITH OTHER SCIENTISTS ON AN
INVESTIGATION OF THE GASES
EXCHANGED DURING BREATHING.

PHOTO: Courtesy of the Hagley Museum
and Library

In other experiments, the Lavoisiers heated elements in enclosed containers with air inside them. They discovered that new substances were formed but that the container and its contents had the same mass as they did before heating. When they measured the mass of the new solid substances formed, they discovered that the new solid substances were heavier than the original elements that were heated. In this way, the Lavoisiers determined that the substances must have gained their mass from the air. Based on these experiments, Antoine Lavoisier also concluded that air contained several gases, one of which reacted with the elements in the experiment. He called this gas oxygen (which had previously been discovered and described— but not named—by Carl Wilhelm Scheele and Joseph Priestley).

In 1789, Antoine Lavoisier wrote the best textbook on chemistry the world had seen. In it, he introduced a new scientific law that he called the law of conservation of mass. This law stated that in any closed system (as small as a sealed container or as big as the whole universe) the total mass remains the same, regardless of what changes take place inside. ■

DISCUSSION QUESTIONS

1. A closed container of air and 5 g of tin metal have a total mass of 765 g. After heating, the container and its contents have the same mass as before (765 g). When the container is opened briefly, a whoosh of air enters it. The container and its contents now have a mass of 766 g. The mass of the tin is now 6 g. How does the mass gained by the tin compare with the mass of air that entered the container when it is opened and why?

2. Imagine you are Antoine or Marie Lavoisier. How could you design an experiment to investigate what happens to the total mass of matter when a caterpillar eats a leaf? Draw a picture of the apparatus you would use, and write a short description of the procedure you would follow. Remember that the Lavoisiers did not have electronic balances.

FINAL ASSESSMENT FOR *Experimenting with Mixtures, Compounds, and Elements*

INTRODUCTION

This lesson is designed to assess how much you have learned while working on the unit *Experimenting with Mixtures, Compounds, and Elements*. The assessment consists of two parts: a performance assessment (Inquiry 13.1) and a written assessment.

▶ IN THIS LESSON, YOU WILL CONDUCT AN INQUIRY THAT REQUIRES YOU TO FOLLOW INSTRUCTIONS CAREFULLY. YOU WILL ALSO HAVE TO MAKE AND RECORD ACCURATE MEASUREMENTS AND OBSERVATIONS.

PHOTO: © Terry G. McCrea, Smithsonian Institution

OBJECTIVES FOR THIS LESSON

Use your knowledge and skills to complete an assessment of what you have learned during the unit *Experimenting with Mixtures, Compounds, and Elements*.

▶ **MATERIALS FOR LESSON 13**

For you

1	copy of Student Sheet 13.1: Performance Assessment
1	copy of Student Sheet 13: Written Assessment
1	pair of safety goggles

For you and your partner

1	jar of solid A
1	bottle of clear solution B
1	250-mL beaker
1	lab scoop
3	test tubes
1	metric ruler
1	black permanent marker
3	labels
	Access to an electronic balance
	Access to water

GETTING STARTED

1 Your teacher will assign a set of materials and a balance to you and your lab partner. Make sure you have all the apparatus listed in the materials list.

2 You may refer to your Student Guide and science notebook for Inquiry 13.1. You may talk to your lab partner, but do not share information or results with other pairs of students. Answer questions on your own.

3 Your teacher will explain when and how to do the written assessment.

SAFETY TIP
Wear your safety goggles throughout Inquiry 13.1.

▶ MAKE SURE TO FILL YOUR TEST TUBES WITH JUST THE AMOUNT OF SUBSTANCE SPECIFIED TO MEASURE MASS ACCURATELY.

PHOTO: Kristen Holden/
creativecommons.org

PERFORMANCE ASSESSMENT

PROCEDURE

1 You have 35 minutes to complete this part of the assessment.

2 In this inquiry, you need to measure mass accurately. Be careful not to spill solids or liquids after you have measured their mass. If you do, start the experiment again. If you make a mistake or spill anything, you may request replacement items from your teacher.

3 Use the black permanent marker and labels to label the three test tubes (A, B, and C) in the beaker.

4 Put one level lab scoop of solid A into test tube A (see Figure 13.1). Stand the test tube in the beaker.

5 Pour water to a depth of about 4 cm into test tube B. Use the ruler to measure the depth. Stand the test tube in the beaker.

6 Pour the clear solution from the bottle labeled B to a depth of about 4 cm in test tube C. Stand the test tube in the beaker.

7 Carry the beaker and test tubes to the balance.

▶ PUT ONE LAB SCOOP OF SOLID A INTO TEST TUBE A.
FIGURE **13.1**

Inquiry 13.1 continued

▶ MEASURE THE MASS OF THE
BEAKER AND THE THREE TUBES
ON THE BALANCE.
FIGURE **13.2**

8 Measure the mass of the apparatus (see Figure 13.2). Record your result in Table 1 (under "Before Mixing Contents of Tube A and Tube B") on Student Sheet 13.1: Performance Assessment.

9 Remove the beaker and test tubes from the balance. Return to your lab station.

10 Without spilling any, pour the water from test tube B into test tube A (see Figure 13.3). Return test tube B to the beaker.

11 Gently shake test tube A from side to side as shown in Figure 13.4. It is very important that you do not place your fingers or thumb over the top of the test tube. Be very careful not to spill any of the liquid. Continue to shake the tube until all the blue crystals dissolve.

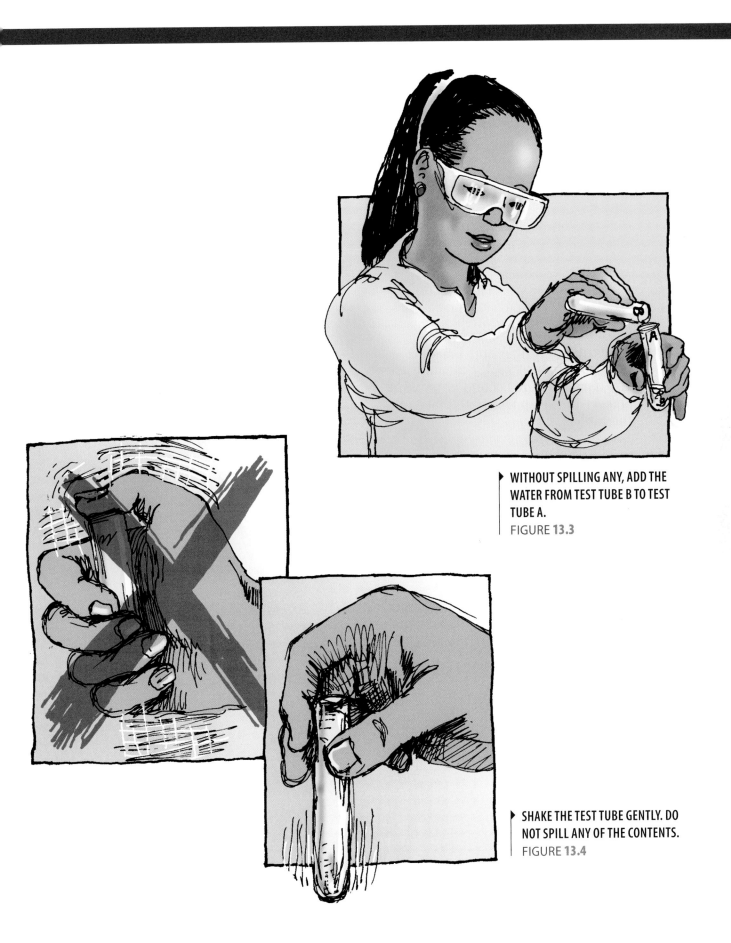

▶ WITHOUT SPILLING ANY, ADD THE
WATER FROM TEST TUBE B TO TEST
TUBE A.
FIGURE **13.3**

▶ SHAKE THE TEST TUBE GENTLY. DO
NOT SPILL ANY OF THE CONTENTS.
FIGURE **13.4**

Inquiry 13.1 continued

12 Return the test tube to the beaker.

13 Remeasure the mass of the beaker and test tubes. Record the mass in Table 1 (under "After Mixing Contents of Tube A and Tube B") on Student Sheet 13.1. Calculate any changes in mass and enter your answer in the appropriate place on the table.

14 Look carefully at the mixture in the test tube. Answer the following question in Step 2 on Student Sheet 13.1: What are three properties of solid A?

15 Carefully pour the contents of test tube C into test tube A. Again, be sure not to spill anything. Return the test tubes to the beaker.

16 Answer the following questions in Step 3 on the Student Sheet 13.1:

A. What did you observe when you mixed the contents of test tube A and test tube C?

B. What evidence was there that a chemical reaction took place?

17 Return to the balance and remeasure the mass of your apparatus. Record the mass in Table 1 (under "After Mixing Contents of Tube A and Tube C") on Student Sheet 13.1. Calculate any changes in mass and enter your answer in the appropriate place on the table.

18 Answer the following questions in Step 4 on Student Sheet 13.1:

A. Was there any change in the mass of the apparatus when you made the solution?

B. Was there any change in the mass after you mixed the two solutions together?

C. Explain these results.

19 Dispose of the contents of your test tubes and rinse them with tap water.

20 Return to your seat and check your answers. Hand in Student Sheet 13.1 before you leave class.

Glossary

acid: A substance that produces or donates hydrogen ions (H⁺) in a solution.

alchemy: An early form of chemical science and philosophy.

alloy: A mixture or solid solution of two metals or a metal and a nonmetal.

base: A substance that produces hydroxide ions (OH⁻) in a solution.

boiling point: The temperature at which a liquid turns into a gas. Boiling points depend on air pressure. Boiling points of substances are usually given for standard air pressure (1 atmosphere).

burning: A rapid chemical reaction between a substance and a gas that produces heat and light. Most burning or combustion takes place in the air and has oxygen as one of its reactants.

characteristic property: An attribute that can be used to help identify a substance. A characteristic property is not affected by the amount or shape of a substance.

chemical property: A characteristic of a substance that involves chemical reactions.

compound: A pure substance consisting of two or more elements combined. For example, sodium chloride consists of the elements sodium and chlorine.

conductor: A substance that allows electricity and/or heat to pass through it.

corrosion: A chemical reaction, usually between a metal and the air. For example, iron reacts with the oxygen in the air, which is called rusting.

density: The mass of a known volume of a substance. It is usually measured in grams per cubic centimeter (g/cm³).

distillation: The process of evaporating a solution and then condensing the various fractions back into a liquid in order to separate them.

electrolysis: The process by which some liquid compounds or some compounds in solution are split into their constituent parts by passing electricity through them.

element: A substance that cannot be broken down into other substances by chemical or physical means (except by nuclear reaction).

filtration: The process of separating a solid and a liquid by passing a mixture of the two through a mesh (usually a filter paper). The liquid, which passes through the filter paper, is called the filtrate. The solid, which remains on the filter paper, is called the residue.

freezing point: The temperature at which a liquid turns to a solid.

gas: A state or phase of matter in which a substance has no definite shape or volume. Oxygen is an example of a gas.

kinetic-molecular theory: Theory that atoms in all forms of matter are in motion: those of gases are the fastest and solids the slowest.

liquid: A state or phase of matter in which a substance has a definite volume but no definite shape. Liquids take the shape of the part of the container they occupy.

magnetic: Property of being attracted to a magnet.

mass: A measure of the amount of matter in an object. Mass is measured in grams or kilograms.

melting point: The temperature at which a solid turns into a liquid. Melting points of substances are altered by changes in pressure and are usually given for standard air pressure (1 atmosphere).

metals: A group of elements that are usually hard solids and that have the following common characteristics: shiny appearance, good conductivity, and malleability. See also *nonmetals.*

mixture: Two or more elements or compounds that are mixed together but are not chemically combined.

nonmetals: A group of elements with characteristic properties that are different from the characteristic properties of metals.

Nonmetals are nonconductive, brittle, and dull in appearance. See also *metals.*

product: A substance formed by a chemical reaction.

pure substance: Either an element or a compound.

reactant: A starting substance in a chemical reaction.

salt: Compound formed from the reaction of a base and an acid.

sedimentation: The process by which a solid settles out of a solid/liquid mixture. The solid that collects below the liquid is called a sediment, or precipitate.

smelting: The process by which a metal is extracted from ore. Smelting involves heating the ore, usually with a source of carbon.

solid: A phase or state of matter in which a substance has a definite shape and volume.

solubility: The amount of solute that will dissolve in a solvent at a given temperature and pressure.

synthesis reaction: A chemical reaction in which the reactants are elements. A compound is made of the two (or more) elements.

thermal decomposition: A chemical reaction in which a compound is decomposed by heating.

volume: The amount of space occupied by a sample of matter. Volume is measured in liters (L) and milliliters (mL) as well as in cubic centimeters (cm^3) and cubic meters (m^3).

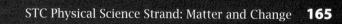

Index

Z
Zinc, 81
    conductivity of, 6
    corrosion of, 125-127
    reactivity with acid, 122-123
Zinc oxide, 26, 81

Photo Credits

Front Cover
NASA Marshall Space Flight Center

Lessons
2 T.K. Ives, Jr. **4** Courtesy of Bristol-Myers Squibb/Charlotte Raymond, Photographer **14** © Carolina Biological Supply Company, used with permission **16** Stephen Ausmus, Agricultural Research Service/U.S. Department of Agriculture **18** U.S. Air Force photo/Staff Sgt. Bennie J. Davis III **19 (top)** Bugeater/creativecommons.org **(bottom)** © David Marsland **20 (top left)** Library of Congress, Prints & Photographs Division, LC-B2-2278-7 **(bottom right)** National Science Resources Center **21** Courtesy of U.S. Navy Blue Angels **22** David Weekly/ creativecommons.org **24** National Science Resources Center **27** NASA African Monsoon Multidisciplinary Analyses (NAMMA) **28 (top)** Ellsworth Air Force Base photo by Airman Nathan Riley **(bottom)** Wikimedia Commons **30 (top)** pearlbear/creativecommons.org **(bottom)** Library of Congress, Prints & Photographs Division, LC-USZ62-70599 **31** Andy Pernick/U.S. Bureau of Reclamation **32** Courtesy of Carolina Biological Supply Company **39** National Science Resources Center **40** Stephen Ausmus, Agricultural Research Service/U.S. Department of Agriculture **42** Joi Ito/creativecommons. org **45** Midtown Crossing at Turner Park/ creativecommons.org **47 (top)** © David Marsland **(bottom left)** William Billard/ creativecommons.org **(bottom right)** © David Marsland **48** © 2009 Carolina Biological Supply Company **49 (left)** © 2009 Carolina Biological Supply Company **(right)** © 2009 Carolina Biological Supply Company **53** Library of Congress, Prints & Photographs Division, LC-USZC4-8658 **54** © Kenji Mishina **56** Library of Congress, Prints & Photographs Division, FSA/OWI Collection, LC-USF 33-016109-M3 **57** Library of Congress, Prints & Photographs Division, FSA/OWI Collection, LC-USF 33-031141-M4 **58** NASA **60** Randolph Femmer/National Biological Information Infrastructure **61 (left)** © 2009 Carolina Biological Supply Company **(right)** © 2009 Carolina Biological Supply Company **62 (top right)** © 2009 Carolina Biological Supply Company **(bottom left)** © 2009 Carolina Biological Supply Company **63** © 2009 Carolina Biological Supply Company **66** DoD photo by Cpl. Bryson K. Jones, U.S. Marine Corps **67** U.S. Air Force photo by Senior Master Sgt. David H. Lipp **68** © David Marsland **71** National Air and Space Museum, Smithsonian Institution (SI 92-3598) **72** National Air and Space Museum, Smithsonian Institution (SI 77-15140) **73** National Air and Space Museum, Smithsonian Institution (SI 76-3577) **75 (top right)** National Museum of American History, Smithsonian Institution (SI 99-2991) **(left)** Saquan Stimpson/creativecommons. org **76 (top right)** The reproduction of this image is through the courtesy of Alcoa Inc. **(bottom right)** The reproduction of this image is through the courtesy of Alcoa Inc. **(bottom left)** The reproduction of this image is through the courtesy of Alcoa Inc. **77** The reproduction of this image is through the courtesy of Alcoa Inc. **78** Library of